普通高等学校双语教学规划教材

# Experimental College Physics

主编　武颖丽　李平舟
审译　侯娟娟

西安电子科技大学出版社

## 内容简介

本书是专门为外国留学生、合作办学学生和准备出国留学的学生编写的物理实验教材，所选实验内容紧扣大学物理教材，介绍了常用物理量的测量、常用仪器的结构及使用，涵盖测量方法、误差理论、数据处理、力学、电学、光学等内容。全书共编入21个经典实验，其中力学5个、光学7个、电学9个。

本书可作为高等院校各专业的物理实验双语教材，也可作为教师或科技人员的参考书。

**图书在版编目(CIP)数据**

**大学物理实验**/武颖丽，李平舟主编. —西安：西安电子科技大学出版社，2016.11
普通高等学校双语教学规划教材
ISBN 978-7-5606-4290-1

Ⅰ.① 基… Ⅱ.① 武… ② 李… Ⅲ.① 物理学—实验—双语教学—高等学校—教材
Ⅳ.① O4-33

**中国版本图书馆CIP数据核字(2016)第224161号**

策　　划　云立实
责任编辑　周正履　云立实
出版发行　西安电子科技大学出版社(西安市太白南路2号)
电　　话　(029)88242885　88201467　　邮　　编　710071
网　　址　www.xduph.com　　电子邮箱　xdupfxb001@163.com
经　　销　新华书店
印刷单位　陕西天意印务有限责任公司
版　　次　2016年11月第1版　2016年11月第1次印刷
开　　本　787毫米×960毫米　1/16　印张10.5
字　　数　212千字
印　　数　1～2000册
定　　价　22.00元
ISBN 978-7-5606-4290-1/O
**XDUP 4582001-1**

# 前言

"大学物理实验"是高等工科院校课程体系中一门重要的基础课程，也是大学生进校后的第一门科学实验课程。该课程旨在为后续专业实验和科学研究奠定基础，不仅要讲授学生开展实验工作所需具备的基本知识、基本方法和基本技能，更要培养学生观察问题、分析问题和解决问题的科学思维与创新意识，以及理论联系实际的能力。

随着国际交流日益频繁，高等教育国际化是不可扭转的趋势。为适应国际化教育发展的需要，在西安电子科技大学物理实验中心、教务处和出版社的大力支持下，作者以中文版《物理实验》(西安电子科技大学出版社，2007)为基础，结合多届外国留学生的教学经验，编写了本书。本书的编写已被纳入西安电子科技大学"十三五"规划教材建设。

大学物理实验是集体教学活动，本书凝结了物理实验中心全体教师的集体智慧和辛勤劳动。本书编写过程中得到了许多教师的大力支持，同时参考了国内外其他高等学校的同类教材，在此一并表示衷心感谢！还要感谢本书的审译——延安大学外国语学院的侯娟娟老师付出的辛勤劳动。

实验教学探索是永无止境的任务，加之编写时间仓促及编者业务水平有限，书中难免有不妥与疏漏之处，恳请同行及广大读者提出宝贵意见。

编　者

2016年5月

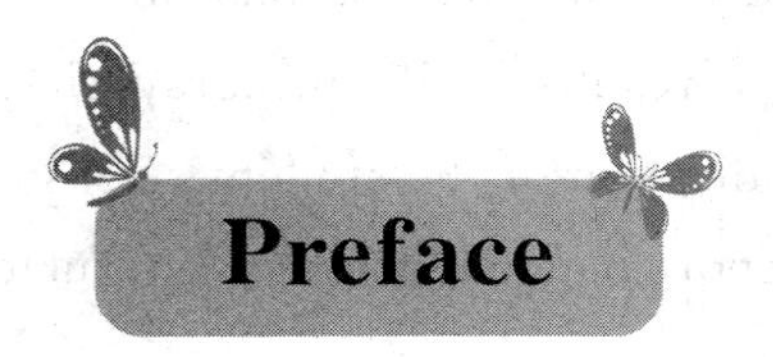


# Preface

Experimental College Physics has 21 experiments including mechanics, electricity and optics. It aims to provide "hand-on" experiences of various physical principles and to augment and supplement the learning and understanding of basic physical principles, while introducing laboratory procedures, techniques, and apparatus. In so doing, students become familiar with laboratory equipment, procedures and related scientific methods. In this book, the theory of physical principles is presented in experiments, and the predicted result will be tested by experimental measurements. Even those well-known principles, which may have been tested many times before, are included within some accepted theoretical or measured values. But to be best, you should imagine that you are the first person to perform an experiment to test a scientific theory.

Basically, the textbook is designed for students who are taking their first course in physics, and the order of the subject matter of the experiments is approximately that found in most standard textbooks for first-year college physics. Moreover, the apparatus in all the experiments is of simple design and could be found in most physics laboratories practically. Authors believe that the fundamental principles of physics can best be learned through the use of simple apparatus. If high precision is required in a first course, much of the understanding of fundamental principles is sacrificed to acquiring skill in operating complex equipment.

In addition to the list of apparatus, the instructions for each experiment include a statement of the purpose of the experiment, an introduction summarizing the physical principles involved, and directions for the experimental procedures. A description of the operation and use of the apparatus is included in some of the experiments where such an explanation seems necessary. Questions which follow some experiments are designed to aid students in making more careful observations and to train them to analyze these observations and interpret the results. The authors believe that the answers to these questions give a very clear indication of the student's grasp of the experiment, and are a very important part of the report handed in to the instructor.

## Part Ⅰ Introduction to Experimental Uncertainties and Error Analysis

1. No Information without Uncertainty Estimation ······ ( 3 )

2. What is an Error Bar? ······ ( 4 )

3. Random Errors, Systematic Errors, and Mistakes ······ ( 5 )

4. How to Estimate Error Bars in Data ······ ( 6 )

5. Sample Mean, Standard Deviation, and Standard Error ······ ( 8 )

6. How and When to Throw Out Data ······ ( 10 )

7. Combining Unrelated Sources of Error ······ ( 10 )

8. Error Propagation in Calculations: Functions of a Single Measured Quantity ······ ( 12 )

9. Error Propagation in Calculations: Functions of Several Measured Quantities ······ ( 13 )

10. A Reality Check for Error Propagation: Fractional Uncertainty ······ ( 14 )

11. Significant Figures ······ ( 15 )

12. Graphical Representation of Experimental Data ······ ( 17 )

## Part Ⅱ Experiments

Lab 1 Young's Modulus Lab ······ ( 25 )

Lab 2 Standing Waves ······ ( 31 )

Lab 3 Physical Pendulum ······ ( 38 )

Lab 4 Moment of Inertia Measured by Three-Wire Pendulum ······ ( 42 )

Lab 5 The Prism Spectrometer: Dispersion and the Index of Refraction ······ ( 45 )

Lab 6 The Transmission Diffraction Grating: Measuring the Wavelength of Light ······ ( 53 )

Lab 7 Interference of Light ······ ( 59 )

Lab 8 Basic Measurement ······ ( 65 )

Lab 9 Lenses and Simple Lens Systems ······ ( 70 )

Lab 10 Light Polarization—Malus's Law ······ ( 79 )

Lab 11 The Michelson Interferometer ······ ( 85 )

Lab 12 Diffraction and Interference of Plane Light Waves ······ ( 92 )

Lab 13 Construction of an Ammeter and Voltmeter ······ ( 100 )

Lab 14 Electric and Magnetic Forces ······ ( 106 )

Lab 15 Velocity of Sound in Air ······ ( 111 )

Lab 16 The Ballistic Galvanometer ······ ( 118 )

Lab 17 The Magnetic Field of a Solenoid and Ballistic Galvanometer ······ ( 126 )

Lab 18 Linear and Nonlinear Resistors ······ ( 133 )

Lab 19 Hall Effect Experiment ······ ( 136 )

Lab 20 Mapping the Electric Field ······ ( 144 )

Lab 21 The Measurement of Capacitance and High Resistance by Galvanometer ······ ( 150 )

**Appendix** ······ ( 154 )

**References** ······ ( 160 )

# Part Ⅰ

## Introduction to Experimental Uncertainties and Error Analysis

## 1. No Information without Uncertainty Estimation

XIY is about 59 minutes from the south campus of Xidian University by car. You can learn this from the driving directions on Google Maps, and it is a useful piece of information if you are checking out possible travel bargains. But what if you already have reservations out of XIY and need to know when to leave campus for the airport? Then you'd better know that the drive can easily take as little as 45 minutes or as much as an hour and a half, depending on factors you cannot possibly determine in advance.

Similar comments apply to a vast array of numbers we measure, record, and trade back and forth with each other in our everyday life. I will be off the phone in five minutes—or maybe two to eight minutes. We set the oven to 350°, knowing that the actual temperature might be only 330° when the preheat light goes off.

Just about every number in our lives is actually a stand-in for a range of likely values. Another way of putting this is to say that every value comes with an *uncertainty*, or an *error bar*. I will be off the phone in $5 \pm 3$ minutes; the oven temperature is $350 \pm 25$ degrees, and so on. Sometimes a quantity does have zero error bars: I have exactly one brother. More often, though, numbers have error bars and we ignore them only through the ease of familiarity.

When a situation is unfamiliar, though, suddenly it can be very important to ask about the error bars. Without error bars on the travel time to XIY, you may very well miss your flight. Your dorm room may be about three meters wide, but lugging home that used couch will seem pretty dumb if the room is actually 3 meters $\pm 5$ cm! And it is not just numerical values that can have uncertainties attached to them.

Uncertainties—how to think about them, estimate them, minimize them, and talk about them—are a central focus of Physics experiment. We will learn a handful of statistical definitions and methods, but we will concentrate on whether they make sense rather than whether we can justify them rigorously. Our goal is that we ourselves should be able to talk and think reasonably about the experimental situations we encounter in the lab each time.

## 2. What is an Error Bar?

In a laboratory setting, or in any original, quantitative research, we make our research results meaningful to others by carefully keeping track of all the uncertainties that might have an appreciable effect on the final result which is the object of our work. Of course, when we are doing something for the very first time, we do not *know* beforehand what the result is going to be or what factors are going to affect it most strongly. Keeping track of uncertainties is something that has to be done before, during, and after the actual 'data-taking' phase of a good experiment. In fact, the best experimental science is often accomplished in a surprisingly circular process of designing an experiment, performing it, taking a peek at the data analysis, seeing where the uncertainties are creeping in, redesigning the experiment, trying again, and so forth. But a good rule is to estimate and record the uncertainty, or error bar, for every measurement you write down.

What is an error bar and how can you estimate one? An error bar tells you how closely your measured result should be matched by someone else who sets out to measure the same quantity you did. If you record the length of a rod as $95.0 \pm 0.05$ cm, you are stating that another careful measurement of that rod is likely to give a length between 94.95 cm and 95.05 cm. The word "likely" is pretty vague, though. A reasonable standard might be to require an error bar large enough to cover a majority—over 50%—of other measurement results.

However, it is convenient to have some sort of standard definition of an error bar so that we can all look at each other's lab notebooks and quickly understand what is written there. One common convention is to use "one sigma" error bars; these are error bars which tell us that 68% of repeated attempts will fall within the stated range. The 68% figure is not chosen to be weird, but because it is easy to calculate and convenient to work with in the very common situation of something called 'Gaussian statistics.' We will not go into this in detail, but here is one example of how useful this error bar convention can be: for many, many situations, if 68% of repeated attempts are within one error bar of the initial result, 95% will be within two error bars. The essential point here is that your error bars should be large enough to cover a majority, but not necessarily a vast majority, of possible outcomes.

Finally, an error bar estimates how confident you are in your own measurement or result. It represents how well you did in your experimental design and execution, not how well the group at the next bench did, or how well your lab manual was written. Error bars are part of your data and must follow logically from what *you* did and the observations *you* made; anything else is fraudulent data-taking.

## 3. Random Errors, Systematic Errors, and Mistakes

There are three basic categories of experimental issues that students often think of under the heading of experimental error, or uncertainty. These are random errors, systematic errors, and mistakes. In fact, as we will discuss in a minute, mistakes do *not* count as experimental error, so there are in fact only two basic error categories: random and systematic. We can understand them by reconsidering our definition of an error bar from the previous section.

*An error bar tells you how closely your measured result should be matched by someone else who sets out to measure the same quantity you did.* How is this mysterious second experimenter going to measure the same quantity you did? One way would be to carefully read your notes, obtain your equipment, and repeat your very own procedure as closely as possible. On the other hand, the second experimenter could be independent minded and could devise an entirely new but sensible procedure for measuring the quantity you measured. Either way, the two results are not likely to be exactly the same!

### 3.1 Random Errors

Random errors usually result from human and from accidental errors. Accidental errors are brought about by changing experimental conditions that are beyond the control of the experimenter, such as vibrations in the equipment, changes in the humidity, and fluctuating temperatures. Human errors involve such things as miscalculations in analyzing data, the incorrect reading of an instrument, or a personal bias in assuming that particular readings are more reliable than others. By their very nature, random errors cannot be quantified exactly since the magnitude of the random errors and their effect on the experimental values is different for every repetition of the experiment. So statistical methods are usually used to obtain an estimate of the random errors in the experiment.

### 3.2 Systematic Errors

A systematic error is an error that will occur consistently in only one direction each time the experiment is performed, i. e., the value of the measurement will always be greater (or lesser) than the real value. Systematic errors most commonly arise from defects in the instrumentation or from using improper measuring techniques. For example, measuring a distance using the worn end of a meter stick, using an instrument that is not calibrated, or incorrectly neglecting the effects of viscosity, air resistance and friction, are all factors that can result in a systematic shift of the experimental outcome. Although the nature and the magnitude of systematic errors are difficult to predict in practice, attempt should be made to quantify their effect whenever possible. In any experiment, care should be taken to eliminate as many of the systematic and random errors as possible. Proper calibration and adjustment of the equipment will help reduce the systematic errors leaving only the accidental and human errors to cause any spread in the data. Although there are statistical methods that will permit the reduction of random errors, there is little use in reducing the random errors below the limit of the precision of the measuring instrument.

## 4. How to Estimate Error Bars in Data

Since we are not going into Gaussian (let alone other) statistics, our definition of an error bar remains loose enough so that we should not be too concerned over the exact numerical value we assign to error bars in our experiments. However, we do want to base our error bars on experimental reality, so they can be useful in clarifying our data analysis and results in the end. The overall uncertainty of a result tells us how much trust to place in the specifics of the result. Beyond that, however, identifying the major source(s) of the final uncertainty can guide us in spending our time and effort productively, should we wish to redesign the experiment for better results in the future.

So, how do we assign an error bar to a measurement taken in the lab? Several specific but common situations are covered below. The zeroth rule of error estimation, though, is that we should always think about the meaning of an error bar and assign an error bar that makes sense based on that meaning.

One of the simplest sources of uncertainty is the resolution or quoted accuracy of a measuring device. Many lab devices, such as electrical meters and mass balances, have resolutions specified by their manufacturers. These device uncertainties can be read off the device (sometimes on the bottom surface) or in its manual. However, something as simple as a meter stick also has an effective device resolution. If the stick is marked every millimeter, for example, then if an object ends between the 101 and 102 mm marks it is probably unreasonable to expect observers to do any better than choosing which mark is closer. In this way, an object that is truly 101.4 mm long would be measured at 101 mm, while a 101.8 mm object would be recorded as 102 mm long. A reasonable error bar for the device resolution of the meter stick, then, would be $\pm 0.5$ mm. A device resolution uncertainty can be estimated for just about any measurement device by considering its construction and the reliability of a reasonable observer.

Another source of uncertainty, sample variation, becomes important when we measure a phenomenon that just does not quite come out the same every time. In a hypothetical bean sprout study, we conduct the experiment on more than one plant because we suspect there is random variation from one bean sprout to another. Measuring several plants and taking the mean of their heights seems like a natural way to find out something about average bean sprout growth. Just as importantly, though, measuring several plants gives us an idea of how strong the random variation might be and thus how far off the average might still be from the "true" mean. If we measure twenty plants and all twenty are the same height to within a millimeter, we can be fairly certain that we know the average bean sprout height to better than a millimeter (barring systematic errors). On the other hand, if we measure two plants and their heights are 21.00 cm and 22.00 cm, we should be pretty wary of reporting the overall average to be 21.50 cm. In the next section we will develop formulas for quantities called the *standard deviation* and *standard error* that can be used to find random uncertainty in a quantity based on repeated sampling like this.

The error estimation techniques we have just discussed apply primarily to random errors. How can we estimate systematic errors? First we must consider possible causes of systematic error, then estimate reasonably from theoretical knowledge, additional experiments, or prior experience how much effect these causes might have. If we are measuring the length of a metal rod, the length might reasonably depend on temperature. Perhaps the temperature in the room could be as much as three degrees different from standard 'room temperature' definitions. How much shift could that cause in the rod's length? If we have no experience or reference materials to guide us, we could deliberately cool the rod in a refrigerator, measure the new length, and estimate roughly how much length change occurs per degree. This technique of deliberately exaggerating an effect to estimate its significance is often useful in dealing with systematic errors.

There is one more cardinal rule of error sources: "human error" is *never* a legitimate source of error. That phrase is completely uninformative, and should never be used as an insurance or catch-all in discussing an experiment. Humans cause error, of course, but in specific ways that can be described and quantified.

## 5. Sample Mean, Standard Deviation, and Standard Error

In this section we develop formulas to quantify a measurement and its random error, based on taking the measurement repeatedly in what is supposed to be the same way (this is sometimes called sampling). This is probably the most mathematical section of our error analysis discussion, but even here we will give reasons why our formulas are reasonable without actually rigorously deriving them.

Imagine we sample a quantity repeatedly, yielding measurements ($x_1$, $x_2$, …, $x_N$). While we try to make all the measurements identical, random variation shows up in our list, so to estimate an overall result we quite naturally take the mean:

$$\bar{x} = \frac{\sum_{i=1}^{N} x_i}{N} \tag{1.1}$$

Perhaps we have done $N=10$ repetitions. If we kept going to $N=20$ how would the value of $x$ change? What if we kept going even longer? In other words, how much uncertainty is left in our measurement because of our limited sampling of the random variation? To answer this question, it is useful to step back a bit first.

When we want to combine all $N$ measurements into a single representative result $x_{\text{rep}}$, it is easy and natural to take the mean: $x_{\text{rep}} = \bar{x}$. But why is $\bar{x}$, as defined in Equation (1.1), really the best candidate for $x_{\text{rep}}$? It would be nice to come up with some measure of deviation which is minimized, sample-wide, by this choice. Perhaps we should be trying to minimize the distance between the individual data points and $x_{\text{rep}}$.

That is, maybe we should minimize $\sum_{i=1}^{N} |x_i - x_{\text{rep}}|$. This is a nice thought, but it turns out that $x_{\text{rep}} = \bar{x}$ does *not* minimize this particular deviation measure…so this must not be the right deviation measure to think about if we are taking sample means. On the other hand, it turns out that $\sum_{i=1}^{N} (x_i - x_{\text{rep}})^2$ is minimized by taking $x_{\text{rep}} = \bar{x}$. To see this, we can differentiate the expression with respect to $x_{\text{rep}}$ and set the derivative equal to zero:

$$\frac{\mathrm{d}}{\mathrm{d}x_{\text{rep}}}\left(\sum_{i=1}^{N} (x_i - x_{\text{rep}})^2\right) = 0 \qquad x_{\text{rep}} = \frac{\sum_{i=1}^{N} x_i}{N} = \bar{x} \tag{1.2}$$

Indeed, the sample mean is the representative value that minimizes the *sum of the squares of the individual deviations*. So if the sample mean is a good measure of the overall result, something related to this summed-squared deviation should be a good measure of the overall result's uncertainty!

Let us begin by imagining that we take an $(N+1)^{\text{th}}$ measurement. How far from the previous mean is this single, new measurement likely to be? Well, we can use the summed-squared deviation to help us guess, but probably we should divide the sum by $N$ first to turn it into a mean-squared deviation: $\frac{1}{N}\sum_{i=1}^{N}(x_i-\bar{x})^2$. This still is not a good measure of deviation, since it is still squared—if the measurement is a length in centimeters, for example, this thing is in $\text{cm}^2$ so it can not be a deviation. Therefore we will take the square root: $\sqrt{\frac{1}{N}\sum_{i=1}^{N}(x_i-\bar{x})^2}$ is called the *root mean square* deviation, or *r. m. s.* deviation for short, and it is a useful measure of how far from the mean a single measurement is likely to fall. It turns out that, by doing proper Gaussian statistics, one comes up with a slightly more generous (i. e., larger) estimate of individual deviation from the mean. Thus we define a quantity called the *standard deviation*:

$$std.\ \text{deviation} = \sigma = \sqrt{\frac{1}{N-1}\sum_{i=1}^{N}(x_i-\bar{x})^2} \tag{1.3}$$

The standard deviation is used to estimate how far from the mean a single measurement is likely to fall.

Originally, though, we were trying to answer a different question. We wanted to know how far our calculated mean was likely to be from the true, or ideally and infinitely well sampled, mean. *This* is the uncertainty of our final (mean) result, and we call it the ***standard error*** or ***standard deviation of the mean***. If we increase the number of samples $N$, the standard deviation defined in Equation (1. 3) will not in general get smaller. But certainly taking more measurements in our sample ought improve the standard error. Each new measurement we add won't necessarily make $x$ closer to the ideal, but in general we will creep and wander towards the ideal value. Therefore, the standard error is given by:

$$std.\ \text{Err} = \frac{\sigma}{\sqrt{N}} = \frac{\sqrt{\frac{1}{N-1}\sum_{i=1}^{N}(x_i-\bar{x})^2}}{\sqrt{N}} \tag{1.4}$$

To sum up this rather lengthy discussion of repeated trials or samples:

**In the presence of sample variation, the true value of a quantity can often be calculated by taking the mean of *N* repeated trials. In that case, the standard error as defined in Equation (1.4) is a good estimate for the uncertainty of this mean.**

CAUTION: Students are often tempted to take a set of repeated trials and summarize them as a mean value plus or minus the standard deviation, rather than plus or minus the standard error. Perhaps this happens because 'standard deviation' is a more familiar term than 'standard error', and many calculators and software packages have built in functions for standard deviation but not for standard error.

## 6. How and When to Throw Out Data

There are times when it is legitimate to throw out a data point. From time to time, one repetition of an experiment gives a result so completely out of line with the other trials that we know there must have been an unidentified problem of unusual size. When our data look like a tight cluster with one (or two) faraway outliers, it is all right to throw away the outliers just because they are so far from all the rest of the data.

Likewise, when our repetitions produce results that cluster in two different places, we can sometimes think carefully and figure out what we changed or did wrong in half the trials. If we have good reason to think half the data is contaminated and the other half is good, we can throw out the contaminated half.

HOWEVER, it is not legitimate to throw away data points based on a comparison between experimental results and what we *expected* to get, either based on theory or on someone else's reports. This kind of throwing out is habitual and very dangerous, since a habit of this sort will prevent us from ever discovering anything new or surprising on our own. It is easy to fall into the trap of fudging data (or inflating error bars) to match an experiment to theory.

Habits are hard to break. Develop good habits now, and do not adjust your data to match your expectations.

## 7. Combining Unrelated Sources of Error

In most experimental situations, if we look hard enough, there are many different sources of uncertainty. The simplest measurement example we have considered so far is that of finding the room-temperature length of a metal rod. If we use a meter stick with millimeter markings, an uncertainty of ±0.5 mm is associated with the measurement device. We might estimate an uncertainty of ±0.01 mm from unknown temperature variations. An uncertainty of ±0.8 mm might be estimated from doing repeated trials presumably they are different because we have difficulty holding the rod straight against the meter stick each time, or because the rod is slightly longer on one edge than on the other.

What can we report for the overall uncertainty in the length of our simple metal rod? We must somehow come up with a rule for how to combine several uncertainties which are unrelated to each other, but which all influence a single outcome. We could add these uncertainties together, for an overall uncertainty of ±1.31 mm, but this is too pessimistic. Adding the uncertainties assumes a kind of worst-case scenario in which the unrelated error sources all end up producing errors in the same direction. More likely, one cause makes the measurement too small, another makes it too large, etc.

We could simply use the largest single uncertainty and neglect all the others, giving us a length uncertainty of ±0.8 mm. This, however, is too optimistic. Surely the errors do sometimes combine to make the overall result worse than any one contributing factor.

To combine unrelated error sources, we need a way to add them together without neglecting any of them, and without forcing them to be in the same direction as each other (or opposite each other, either). In another area of math we are already familiar with adding things that are not in the same direction as each other: We know how to add together mutually perpendicular vectors. If a vector $a$ is perpendicular to a vector $b$, then the vector sum $c$ has a length given by the Pythagorean theorem $c=\sqrt{a^2+b^2}$. This is how we add together unrelated uncertainties as well. If a measurement has two unrelated sources of uncertainty $\delta_1$ and $\delta_2$, then the overall uncertainty is given by $\delta=\sqrt{\delta_1^2+\delta_2^2}$. The method extends to deal with more than two unrelated error sources as well.

If a quantity has $n$ unrelated (or *independent*) sources of uncertainty ($\delta_1, \delta_2, \cdots, \delta_n$), then the overall uncertainty is given by

$$\delta=\sqrt{\delta_1^2+\delta_2^2+\cdots+\delta_n^2} \tag{1.5}$$

This way of combining independent errors is known as *adding in quadrature*.

Returning to the example of our simple metal rod, we can see that adding our three errors in quadrature gives an overall uncertainty of ±0.9 mm, or ±0.94 mm if we keep one more decimal place. It is also clear that the temperature related uncertainty of ±0.01 mm is completely unimportant compared to the other two. It is often true that one or two error sources are much more important than all the others, and dominate the overall uncertainty of an experimental result.

When this is the case, it is not very important to carry out a heroic error calculation that includes all error sources! Each identified error source should be recorded and estimated, but as soon as it can be clearly labeled as unimportant, it can be dropped from calculations in the interest of time and sanity.

## 8. Error Propagation in Calculations: Functions of a Single Measured Quantity

We have discussed methods for finding the uncertainty for a direct measurement. Often, however, we must do some calculations with our raw data to arrive at the result we are actually interested in. The calculation may be as simple as measuring the diameter of a circle and dividing by two to find its radius… but once we use any measured quantity in a calculation, we have to keep track of the uncertainty in our calculated result due to uncertainty in the original measurement. Keeping track in this way is called ***error propagation***. There is really only one basic formula that governs error propagation, and we will develop it right now.

Let's make this problem general by saying we have a quantity, $x$, which we can measure directly with uncertainty $\delta x$. There is a function $f(x)$ we are interested in knowing. Being uncertain about $x$ will clearly cause some uncertainty in $f$, so we will call this uncertainty $\delta f$.

The uncertainty in $f$ depends on the uncertainty in $x$, but also on the steepness of the function $f$ in the spot where we are evaluating it. We can express "the steepness of the function $f$" in more precise and mathematical terms—it is the function's derivative, $\mathrm{d}f/\mathrm{d}x$. Thus we have an error propagation rule for functions of a single variable:

$$\delta f = \delta x \left| \frac{\mathrm{d}f}{\mathrm{d}x} \right| \tag{1.6}$$

The absolute value signs are there because error bars give the size of errors, not their direction, so all error bars are expressed as positive numbers.

Example: if we measure the diameter of a circle as $d=1.0\pm0.1$ cm, the radius is $r=d/2=0.5$ cm, with an uncertainty of $\delta r=(1/2)\delta d=0.05$ cm. However, the area of that circle is $A=\pi d^2/4=0.79\ \mathrm{cm}^2$, with an uncertainty of $\delta A=\delta d\,|2\pi d/4|=0.16$ cm.

## 9. Error Propagation in Calculations: Functions of Several Measured Quantities

Real life is often not as simple as measuringx and finding $f(x)$. Most interesting things depend on more than one variable. We have to set aside our tidy function $f$ from the previous section and consider another function, $g$, which is now a function of several variables: $g=g(x, y, z, \cdots)$.

To find the value of $g$, we industriously go into the lab and measure all the independent variables $x\pm\delta x$, $y\pm\delta y$, $z\pm\delta z$, etc. We can calculate a value for g, but what is the uncertainty $\delta g$? From our rule above there is a contribution $\delta x\left|\frac{\partial g}{\partial x}\right|$ due to the uncertainty in $x$. (We change the d's to ∂'s to denote the *partial* derivative with respect to $x$, since $g$ is a function of multiple variables.) But there is also a contribution $\delta y\left|\frac{\partial g}{\partial y}\right|$, and a contribution $\delta z\left|\frac{\partial g}{\partial z}\right|$, and so on. How can we combine all these? The key lies in the realization that each uncertainty contribution is unrelated to the others; they are all independent, and not required to be in the same direction or opposite directions. We have already learned how to combine unrelated (independent) errors—they add in quadrature!

Thus we can finally write down the one and only rule of error propagation:

$$\delta g(x, y, z, \cdots)=\sqrt{\left(\delta x\left(\frac{\partial g}{\partial x}\right)^2\right)+\left(\delta y\left(\frac{\partial g}{\partial y}\right)^2\right)+\left(\delta z\left(\frac{\partial g}{\partial z}\right)^2\right)+\cdots} \tag{1.7}$$

Let's take an example in which we wish to calculate the area of a rectangle. We measure the length $l=2.0\pm0.1$ cm and the width $w=1.2\pm0.1$ cm. The area is then $A=lw=2.4\ \text{cm}^2$, but it has an uncertainty:

$$\begin{aligned}\delta A &=\sqrt{\left(\delta l\left(\frac{\partial A}{\partial l}\right)^2\right)+\left(\delta w\left(\frac{\partial A}{\partial w}\right)^2\right)}\\ &=\sqrt{(\delta l(w))^2+(\delta w(l))^2}\\ &=\sqrt{(0.1\text{cm}(1.2\text{cm}))^2+(0.1\text{cm}(2.0\text{cm}))^2}\\ &=0.2\ \text{cm}^2\end{aligned}$$

Notice a few things about the calculation above. First, even though $l$ and $w$ have the same individual uncertainty, they have unequal contributions to the uncertainty of $A$.

Second, we write the final result as $\delta A=0.2\ \text{cm}^2$ rather than $0.2332\cdots\text{cm}^2$, since with an uncertainty in the first decimal place it is a clear waste of space to write down any more. By writing all numbers $\pm$ uncertainties we can afford to be lax about significant digits, but we should not offend common sense with long strings of irrelevant numerals.

## 10. A Reality Check for Error Propagation: Fractional Uncertainty

In the previous section, we presented and at least partially justified Equation (1.6) for the uncertainty in a function of several variables, based on the uncertainties in each of the measured variables. Armed with Equation (1.6), you need no other error propagation formulas but for complicated functions it can be challenging or at least time-consuming to compute the final uncertainty, and it is useful to have some way of anticipating and/or reality-checking the answers you get. For this purpose, one of the most powerful tools for quickly checking error bar results is the concept of *fractional uncertainty*.

The fractional uncertainty in $x$ is simply the name we give to the quantity $(\delta x/x)$. Thinking in terms of fractional uncertainties is very useful, because the fractional uncertainty in many common functions ($\delta f/f$ or $\delta g/g$) is similar to the fractional uncertainty of the variables. While following Equation (1.6) is imperative for getting quantitatively correct uncertainties, considering fractional uncertainty is a much simpler way to see roughly what values those uncertainties should have.

To illustrate the usefulness of fractional uncertainty, consider propagating errors (using Equation (1.6) in several simple and commonly-encountered functions. First, we consider a product of two variables, possibly with a constant coefficient $c$:

$$g(x, y)=cxy$$

$$\delta g=\sqrt{(\delta x)^2(cy)^2+(\delta y)^2(cx)^2}$$

$$\frac{\delta g}{g}=\frac{\delta g}{cxy}=\sqrt{\frac{(\delta x)^2c^2y^2}{c^2x^2y^2}+\frac{(\delta y)^2c^2x^2}{c^2x^2y^2}}=\sqrt{\left(\frac{\delta x}{x}\right)^2+\left(\frac{\delta y}{y}\right)^2}$$

In this case the fractional uncertainty in $g$ due to each variable, $x$ or $y$, is actually equal to the fractional uncertainty in that variable itself. Or consider a slightly less straightforward relationship:

$$f(x, y)=x^2y$$

$$\delta f=\sqrt{(\delta x)^2(2xy)^2+(\delta y)^2(x^2)^2}$$

$$\frac{\delta f}{f}=\frac{\delta f}{(x^2y)}=\sqrt{\frac{(\delta x)^2 4x^2}{x^4y^2}+\frac{(\delta y)^2x^4}{x^4y^2}}=\sqrt{\left(2\frac{\delta x}{x}\right)^2+\left(\frac{\delta y}{y}\right)^2}$$

Here the fractional uncertainty in $f$ due to $x$ is not quite equal to the fractional uncertainty in $x$. However, it is still comparable, and the relationship between them is a simple one. Many functions we encounter in nature are products and low-order polynomials of this sort; for them, comparing fractional uncertainties in functions and their variables can be a good way to arrive quickly at a roughly correct error propagation result.

However, keep in mind that for some functions Equation(1.6) does indeed lead to fractional uncertainties in functions which are not at all similar to the fractional uncertainties in the variables. For $x = 0.05 \pm 0.01$ radians, what are the value and the uncertainty of $\cos(x)$? For $x = 3.5 \pm 0.1$ cm and $y = 3.4 \pm 0.1$ cm, what are the value and uncertainty of $x - y$? As a final cautionary note, for $x = 10 \pm 1$ find the value and uncertainty in the function $e^x$.

## 11. Significant Figures

The digits required to express a number to the same accuracy as the measurement it represents are known as significant figures. If the length of a cylinder is measured as 20.64 cm, this quantity is said to be measured to four significant figures. If written as 0.0002064 kilometers, we still have only four significant figures. The zeros preceding the 2 are used only to indicate the position of the decimal point. The zero between the 2 and 6 is a significant figure, but the other zeros are not. If the above measurement is made with a meter stick, the last digit recorded is an estimated figure representing a fractional part of a millimeter division. *All recorded data should include the last estimated figure in the result, even though it may be zero.* If this measurement had appeared to be exactly 20 cm, it should have been recorded as 20.00 cm, since lengths can be estimated by means of this instrument, to about 0.01 cm. When the measurement is written as 20 cm it indicates that the value is known to be somewhere between 19.5 cm and 20.5 cm, whereas the value is actually known to be between 19.995 cm and 20.005 cm.

By referring again to the 20.64 cm measurement, the possible error in this measurement is $\pm 0.005$ cm and was recorded as being nearer to 20.64 cm than to 20.63 or 20.65. Hence, the error is less than one part in two thousand.

Now suppose the diameter of the cylinder is measured with the same instrument and recorded as 2.25 cm. This number has only three significant figures, and hence is known to only one part in a little more than two hundred. From this we see that the number of decimal places does not indicate the precision of the measurement.

Now suppose we wish to find the volume of this cylinder as given by the relation $V=\pi r^2 h$. The radius $r=1.125$ cm, four significant figures being retained because the original number has been reduced by the process of dividing by 2, thus giving rise to a larger percentage error from deviations in the third significant figure. The accuracy of the original measurement will determine when it is best to include an additional figure in such cases.

If we underline the doubtful figures in the number representing $r$ and $r^2$, the multiplication is as shown below.

$$
\begin{array}{r}
1.1\underline{25} \\
1.1\underline{25} \\
\hline
\underline{5625} \\
\underline{2250}\phantom{0} \\
11\underline{25}\phantom{00} \\
11\underline{25}\phantom{000} \\
\hline\hline
1.2\underline{65625}
\end{array}
\qquad\qquad
\begin{array}{r}
20.6\underline{4} \\
1.2\underline{7} \\
\hline
1\underline{4448} \\
41\underline{28}\phantom{0} \\
206\underline{4}\phantom{00} \\
\hline\hline
26.\underline{2128}
\end{array}
$$

The result is shown to be 1.265625; but if the doubtful figures are carried through the process of multiplication and only one of them kept in the final result, the value of $r^2$ is recorded as $1.2\underline{7}$.

If the first 6 in the result is doubtful, the other four figures are worthless in the result and should be discarded. In like manner the product $r^2 h$ has a value of 26.2 when we include only one doubtful figure. It should be noted that this final product contains no more significant figures than does the factor having the fewest significant, namely, 1.27, which has three.

The next step is to multiply by $\pi$, the value of which that you have most probably been using is 3.1416. This multiplication is being left as an exercise for the students under the supervision of the instructor and supplemented by his discussion.

First, multiply the result of $r^2 h$ as given above by 3.1416, showing all the steps in the multiplication and indication the doubtful figures, and record the final result so as to retain only one doubtful figure. Now multiply the value of $r^2 h$ by 3.14 and record the final result as containing one doubtful figure. How do the two results compare? What rule would you suggest concerning the number of digits to use for $\pi$ in multiplication processes such as this? If the diameter of a certain circle is 9.81 cm, with only one doubtful figure, what should be used as the value of $\pi$ in obtaining the circumference of the circle? Check the validity of your answer by multiplying 9.81 by 3.14, then by 3.142, then finally by 3.1416. If you keep only one doubtful in the final result, how many significant figures of $\pi$

are required, and how many significant figures are in your answer? Note that 9.81 is almost as large as 10.00, a number having four significant figures. Hence, one must carry enough digits in $\pi$ to avoid introducing more uncertainty into the answer.

Now take the diameter of the cylinder as 3.28 cm instead of 2.25 cm and calculate the volume by following the doubtful figures through the multiplication process. If one doubtful figure is retained in the result, how many significant figures appear in your answer? Is this number any different from what you expected?

Your result from the first calculation of the volume should be 82.3 $cm^3$. Now suppose the cylinder weighed 784.7 g and we wished to find the weight per unit volume. We should find the quotient of $784.7 \div 82.3$. By adding sufficient zeros to carry out the division in the usual manner we obtain 9.534629 $g/cm^3$ as the weight per unit volume. If the effect of the doubtful figures in both numbers is followed through, it should be noted that the number 9.53 has as many significant figures as we are justified in keeping.

***General rules for computations with experimental data.***

ⅰ. In addition and subtraction, do not carry the result beyond the first column which contains a doubtful figure.

ⅱ. In multiplication and division, carry the result to the same number of significant figures that there are in that quantity entering into calculation which has the least number of significant figures. If the first digit of this quantity is 7 or more, than for safety purpose, it should be considered as having one additional significant value.

ⅲ. In dropping the figures which are not significant, the last retained should be increased by 1 if the first dropped is 5 or more.

## 12. Graphical Representation of Experimental Data

From an examination of the tabulated values of a number of measurements of related quantities it is often difficult to grasp the relationship existing between the numbers. A method widely used to discover such relationships is the graphical method, which gives a pictorial view of the results and makes it possible to interpret the data by a quick glance.

① *Independent and dependent variables.* In any experimental study of cause and effect the aim is to vary just one condition at a time (the cause) and to observe the corresponding values of another quantity (the effect) to see its relation to the first condition. The existing relationship is most easily interpreted from the graph, with the independent variable plotted on the abscissa scale ($X$-axis) and the dependent variable, on the ordinate scale ($Y$-axis). Very often the values to be plotted are all positive and only the first quadrant of a rectangular coordinate system will be needed. In such cases the origin should be shifted to the lower left-hand corner of the sheet of cross section paper. When possible draw the axes inside the margins of the graph paper along the first or second large square. This gives more space to write in the scale and also furnishes guide lines for lettering the names of the variables being plotted (Fig. 1. 1). Graph paper with 20 squares to the inch is recommended for the curves to be plotted in this course.

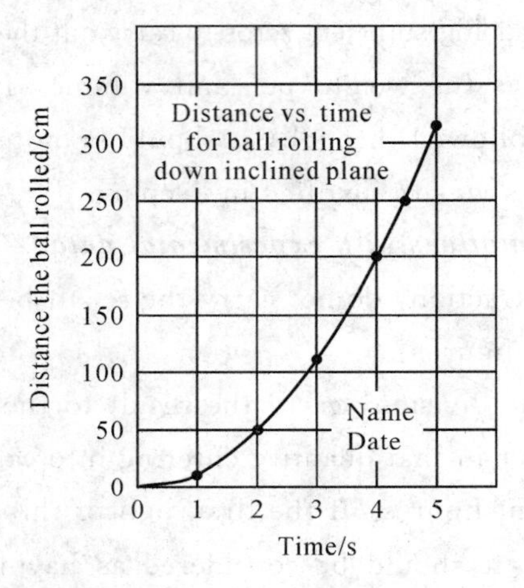


Fig. 1. 1 A sample graph showing the general form that should be used for curve plotting in general.

② *Choice of scale*. Choose a size of graph that bears some relation to the accuracy with which the plotted data are known. In general, the curve should fill most of the sheet if the data are known to three significant figures. If the data are known to only two significant figures, a large graph gives a false impression of the precision of the measurements.

Note the range of values of the independent variable ($X$ quantity), and number of spaces along the $X$-axis. Choose a scale for the main divisions on the graph paper that are easily subdivided so that all the range of values may be included. Subdivision such as 1, 2, 5, and 10 are best, but 4 is sometimes used; 3, 7, or 9 are never used, since these make it very difficult to read values from the graph. The same procedure should be used for the ordinate scale, but the divisions on the ordinate and abscissa scales need not be alike. In many cases it is not necessary that the intersection of the two axes represent the zero values of both variables. If the values to be plotted are exceptionally large or small, use some multiplying factor that permits using a maximum of two or three digits to indicate the value of the main division. A multiplying factor such as $\times 10^{-2}$ or $\times 10^{-6}$ placed on the right of the largest value on the scale may be used.

③ *Labelling*. After you have decided which variable is to be plotted on which axis, neatly letter the name of the quantity being plotted together with the proper unit (see Fig. 1.1). Abbreviate all units in standard form. Then write the numbers along main divisions on the graph paper, using an appropriate scale as explained in the preceding paragraph. The title should be neatly lettered on the body of the graph paper, but it is usually best to do this after the points have been plotted so the title will not interfere with the curve. Explanatory legends should also be shown.

④ *Plotting and drawing the curve*. Use a sharp pencil and make small dots to locate the points. Do not write the coordinates of the points on the graph paper. Your table of data shows these. Carefully encircle each point with a circle about 2 or 3 mm in diameter. In drawing the graph it is not always possible to make all of the points lie on a smooth curve. In such cases, a smooth curve should be drawn through the series of points so as to follow the general trend and thus represent an average. Suppose the plotted points show a straight line trend as shown in Figs. 1.2 and 1.3. To draw the straight line which best represents the relationship which produces the series of points, proceed as follows. First, cover the lower half of the points and draw a faint sharp cross in the centroid of the points in the top half the series. Next, cover the top half and mark a cross at the centroid of the lower group. Then draw a line straight through the two crosses and it will represent a true average. If the series of points appear to represent a function which is not a straight line, the points should lie on both sides of the curve along *all* parts of the curve.

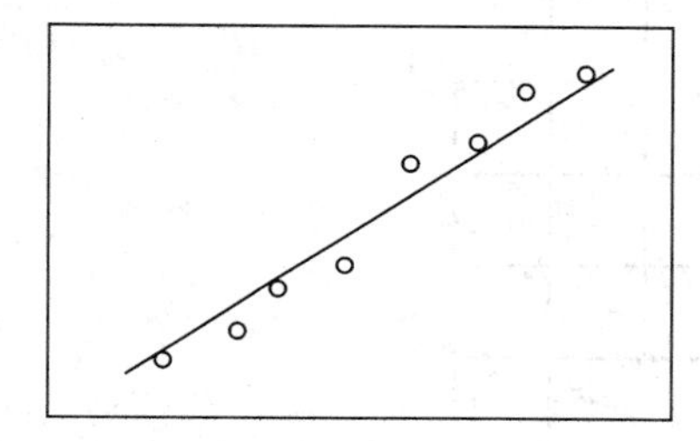

Fig. 1.2 Incorrect method of fitting a straight line to a series of points.

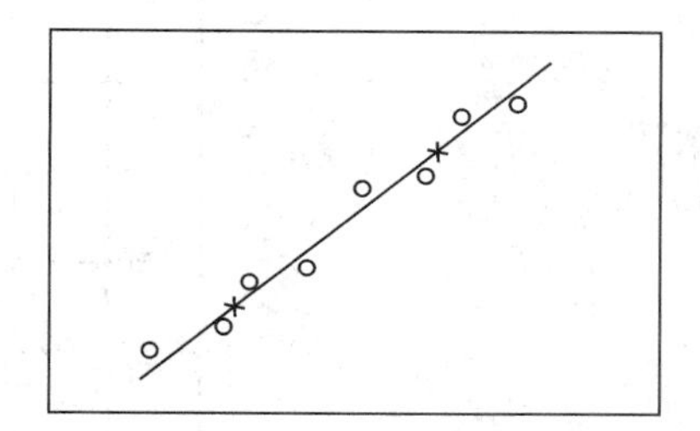

Fig. 1.3 Correct method of fitting a straight line to a series of points.

When more than one curve is drawn on a graph, and it is desirable to distinguish the points associated with one curve from those associated with another, crosses (╳), triangles (△), squares (□) and circles (○) may be used.

⑤ *Analysis and interpretation of graphs.* One of the principal advantages afforded by graphical representation is the simplicity with which new information can be obtained directly from the graph by observing its shape and intercepts.

⑥ *The shape of a graph* immediately tells one whether the dependent variable increases or decreases with an increase of the independent variable. It also shows the rate of change. If the points lie along a straight line there is a linear relationship between the variables. If the variables are directly proportional to each other, they approach zero simultaneously and the line passes through the origin. Curves which are straight lines and do not pass through the origin do not indicate direct proportion.

⑦ *Slope.* In discussing the slope of a graph we must distinguish between physical slope and geometric slope. The geometric slope is usually the angular inclination of the line with respect to the $X$-axis. In plotting physical data there may be an enormous difference in the size of the units on the two axes. The physical slope is found by dividing $\Delta y$ by $\Delta x$ (see Fig. 1.4), using for each the scales and units that have been chosen for those axes. The unit of the slope will be the ratio of the units on the respective axes. With physical slope, one is not usually concerned at all with the angle the line makes with the $X$-axis; the tangent of the angle has no meaning.

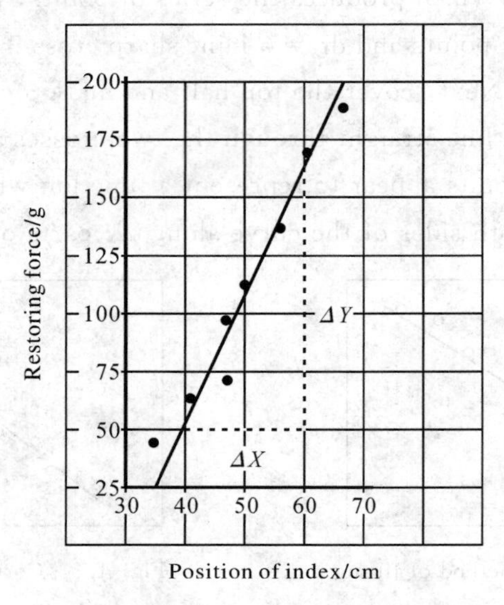


Fig. 1.4 Analysis for stiffness constant of spring.

⑧ *Intercepts.* Significant information is often revealed by the intersections of the graph with the coordinate axes, which is true for other types of curves as well as for straight lines. The true interpretation of the intercept can be obtained only if the scales used begin at zero. In many cases there are no data available for drawing the curve to the axes. If the plotted points indicate the trend of the curve, one may be justified in.

⑨ *Extrapolating* the curve to the intercept desired. *Extrapolation* is accomplished by extending the curve in the desired direction by a dotted line, rather than a solid line, indicating that data are not available for this portion of the curve. Intercepts obtained by extrapolation may serve to aid in the theoretical interpretation of the phenomena being observed.

# Part Ⅱ

## Experiments

# Lab 1 Young's Modulus Lab

In most applications and example problems used in introductory physics courses, objects are assumed to be rigid for the purpose of simplification. Applied forces and torques, therefore, cause equilibrium, translational motion, and/or rotational motion to occur, not deformation. Personal experience, however, tells us otherwise. When supposedly rigid materials are subject to great forces, permanent deformation is a definite possibility. Car crashes are an unfortunately common example. Stories of buildings and bridges collapsing under duress are less common, but they do occur.

## Purpose

1. To determine the value of Young's Modulus for a wire made of an unknown metal;
2. To predict the type of metal used in the making of the wire;
3. To develop skills in graphing and doing calculations.

## Apparatus

Young's Modulus measuring apparatus; telescope; optical lever; micrometer gauge; vernier calipers; steel tape; weights.

## Theory

Engineers need to be extremely cognizant of the properties of the materials that they use in their designs. When subject to a particular *stress*, or force per unit area, materials will respond with a particular *strain*, or deformation. If the stress is small enough, the material will return to its original shape after the stress is removed, exhibiting its elasticity. If the stress is greater, the material may be incapable of returning to its original shape, causing it to be permanently deformed. At some even greater value of stress, the material will break or fracture. The particular values of stress that cause these three situations differ for every material. Knowing these values, however, is vitally important.

There are different types of stress: tension or tensile stress, compression or compressive stress, shear stress, and hydraulic stress. The quantity for all types of stress, however, can be defined as follows:

$$\text{stress} = \frac{\text{force}}{\text{area}} = \frac{F}{A} \tag{1.1}$$

where $F$ is the force applied and $A$ is the cross-sectional area of the material. (In the case of hydraulic stress, $A$ represents the surface area of the material.) Notice that the standard unit of stress is $N/m^2$.

When stress is applied to a body it deforms in some fashion. This deformation is measured by the strain which is defined uniquely for different types of deformation. In general it is some deformation per unit dimension. For a wire supporting a weight the deformation is the elongation and the strain is defined as the elongation per unit length. The quantity strain (for tensile and compressive types of stress) can be defined as follows:

$$\text{strain} = \frac{\Delta L}{L} \tag{1.2}$$

where $L$ is the original length of the material, and $\Delta L$ is the change in length that results after the stress is applied. Notice that strain is a dimensionless quantity. (In the cases of shear and hydraulic stress, strain is defined slightly differently.)

Young's Modulus, $E$, is a constant that describes the ratio of stress to strain for a material experiencing either tensile or compressive stress.

$$E = \frac{\text{stress}}{\text{strain}} \quad \text{or} \quad \frac{F}{A} = E\frac{\Delta L}{L} \tag{1.3}$$

In the case of tensile stress on a cylinder or wire, the following diagram illustrates the above variables as Fig. 1.1.

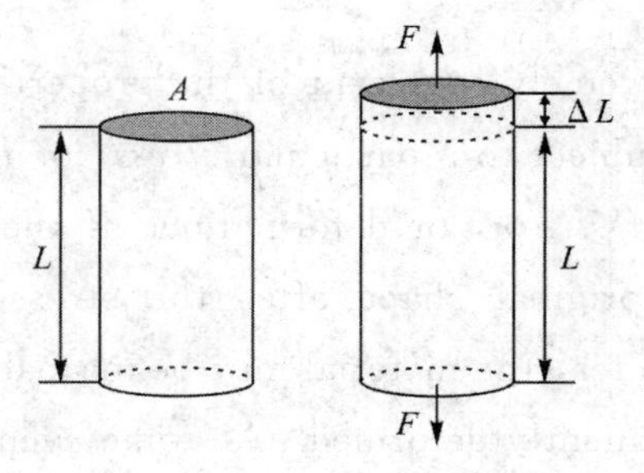


Fig. 1.1 Tensile stress on a cylinder or wire.

In the case of compressive stress, the forces act in the opposite direction, causing the cylinder to compress by a length $\Delta L$. (The definitions of the shear modulus, G, and the bulk modulus, $B$, can be found in other references)

Young's Modulus for many materials is available in handbooks. The modulus is then known as Young's Modulus $E$ and hence

$$E=\frac{F \cdot L}{A \cdot \Delta L} \tag{1.4}$$

where $F$ is the force in N; $A$ is the cross sectional area in $m^2$; $\Delta L$ is the increase in length (in m) caused by $F$; and $L$ is the original length of the wire in m.

In this experiment we use an optical lever to measure the small elongations. Optical lever is made up of one small circular plane mirror and three metal toes which form one isosceles triangle. The length of the back metal toe can be adjusted if necessary. The distance $b$ from the back metal toe to the line between two front metal toes is called the length of optical lever.

See Fig. 1. 2. The normal of plane mirror is accordant with the axes of telescope when there is no force on the wire. Now the reading of measuring scale from the telescope is $N_0$. When adding the weight, the wire is stretched $L$; the back metal toe of the optical lever is dropped so that the angle $\alpha$ of the normal of the plane mirror is formed. At this time, the reading of measuring scale from the telescope is $N_2$. According to the reflection law of light, $\angle N_0ON_1=\angle N_1ON_2$ and $\angle N_0ON_2=2\alpha$, where $N_1$ is the position of the normal of plane mirror in latter situation. The difference of reading of measuring scale between the

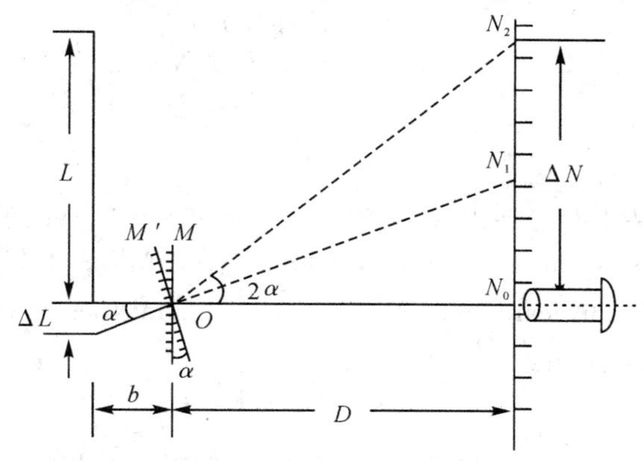


Fig. 1. 2 The operation of optical lever.

initial and latter situations is $\Delta N = |N_2 - N_0|$.

According to geometry, we know

$$\tan\alpha = \frac{\Delta L}{b} \tag{1.5}$$

$$\tan 2\alpha = \frac{\Delta N}{D} \tag{1.6}$$

where $D$ is the distance between the measuring scale and the telescope, $D = ON_0$.

Because $L$ is a small quantity and $\Delta L \ll b$,

$$\tan\alpha \approx \alpha \approx \frac{\Delta L}{b} \tag{1.7}$$

Similarly, $\Delta N \ll D$,

$$\tan 2\alpha \approx 2\alpha \approx \frac{\Delta N}{D} \tag{1.8}$$

Then

$$\frac{\Delta L}{b} = \frac{\Delta N}{2D} \tag{1.9}$$

Namely,

$$\Delta N = \frac{2D}{b} \cdot \Delta L = K \cdot \Delta L \tag{1.10}$$

where $K$ is the magnification for the optical lever, $K = 2D/b$.

Finally, substitute Equation (1.10) and $A = \pi d^2/4$ into Equation (1.4). If $mg$ represents the weight of a mass applied to the wire, then, the Young's Modulus can be written as

$$E = \frac{FL}{A\Delta L} = \frac{8mgLD}{\pi d^2 b}\frac{1}{\Delta N} \tag{1.11}$$

## Procedure

At the beginning of the experiment, examine your apparatus very carefully and work with it until you understand how to measure elongation. Adjust the height of the telescope to the same height of the optical lever. The two front metal toes of the optical lever are put in the transverse groove of the flat while the back metal toe of the optical lever is put on the clamp.

(1) Move the telescope to point to the plane mirror by making the gap and post on the telescope and the plane mirror in one line.

(2) Put the eye outside the telescope in the direction of gap-post; look for the image of the measuring scales from the outside of the telescope; if you can not find the object, adjust the telescope position and the angle of the plane mirror.

(3) Put the eye on the telescope; adjust the ocular to make the cross clear; adjust the focal length of the telescope to make the image of the measuring scale clear; and record the reading on the gauge of horizontal when no weight is on the weight holder, "$N_0$".

(4) Add weights to the weight holder; the telescope will give readings of the measuring scale. Take the telescope reading and find the original length of the wire from its support to the point of attachment to the clamp. Load the measured wire successive kilogram weights, and take the telescope reading after each addition. Unload one weight per time, and read the telescope at each step.

(5) Measure the diameter of the wire at three points along its length in two directions and find a mean value. Measurement must be done carefully with a micrometer screw caliper.

## Record and Calculation

1. Fill in the table and calculate the difference of the telescope readings.

**Table 1.1 The data for telescope reading.**

<table>
<tr><th rowspan="2">Number<br>$i$</th><th rowspan="2">Weights<br>$m$/kg</th><th colspan="3">Telescope reading/mm</th><th rowspan="2">Difference of reading<br>$\delta N = N_{i+4} - N_i$</th><th rowspan="2">difference<br>$\lvert \delta(\delta N) \rvert$</th></tr>
<tr><th>loading</th><th>unloading</th><th>Mean $N_i$</th></tr>
<tr><td>0</td><td></td><td></td><td></td><td></td><td></td><td></td></tr>
<tr><td>1</td><td></td><td></td><td></td><td></td><td></td><td></td></tr>
<tr><td>2</td><td></td><td></td><td></td><td></td><td></td><td></td></tr>
<tr><td>3</td><td></td><td></td><td></td><td></td><td></td><td></td></tr>
<tr><td>4</td><td></td><td></td><td></td><td></td><td rowspan="4">$\overline{\delta N} =$</td><td rowspan="4">$\overline{\delta(\delta N)} =$</td></tr>
<tr><td>5</td><td></td><td></td><td></td><td></td></tr>
<tr><td>6</td><td></td><td></td><td></td><td></td></tr>
<tr><td>7</td><td></td><td></td><td></td><td></td></tr>
</table>

2. Measure and record the data for the radius of the wire.

**Table 1.2 The data for the radius of the wire.**

| Number | 1 | 2 | 3 | 4 | 5 | 6 | Mean $\bar{d}$ |
|---|---|---|---|---|---|---|---|
| $d$/mm | | | | | | | |

3. Measure and record data for the length of the back leg, length of the wire, distance between the telescope and light lever.

**Table 1.3 The data for the length of back leg, length of the wire, distance between the telescope and light.**

| $b$/mm | $\Delta b$/mm | $L$/mm | $\Delta L$/mm | $D$/mm | $\Delta D$/mm |
|---|---|---|---|---|---|
| | | | | | |

4. Calculate Young's Modulus E and its uncertainty.

$$E=\frac{8mgLD}{\pi d^2 b}\frac{1}{\Delta N}=\_\_\_\_\_\_\_\_\_\_\quad (\text{Unit:}\qquad\qquad)$$

$$E_E=\frac{u_E}{E}=\sqrt{\left(\frac{u_b}{b}\right)^2+\left(\frac{u_L}{L}\right)^2+\left(\frac{u_D}{D}\right)^2+\left(\frac{u_d}{d}\right)^2+\left(\frac{u_{\Delta N}}{\Delta N}\right)^2}=$$

$$u_E=E_E\cdot E=(\_\_\_\_\_\_\_\_\_\_)$$

$$E=\bar{E}\pm\overline{\Delta E}=(\_\_\_\_\_\_\_\_\_\_\pm\_\_\_\_\_\_\_\_\_\_)\mathrm{N/m^2}$$

## Questions

(1) Why should the distance $D$ be so long?

(2) We can plot the figure about the stress versus the strain. If the stress versus strain graph for a different material had a steeper slope, what would this suggest?

(3) What are the definitions of stress, strain and elastic limit?

# Lab 2 Standing Waves

## Purpose

1. To observe resonant vibration modes on a string;

2. To determine the frequencies with number of nodes, tension, length and string density;

3. To determine the velocity of transverse waves in the string.

## Apparatus

Variable frequency oscillator; pulley and weights; string; meter stick; scale.

## Theory

Have you ever wondered why pressing different positions on your guitar string produces different pitches or sounds? Or why the same sound is produced by pressing certain positions on two or more strings? By exploring several basic properties of standing waves, you will be able to answer some of these questions. In this experiment, you will study standing waves on a string and discover how different modes of vibration depend on the frequency, as well as how the wave speed depends on the tension in the string.

1. Formation of standing waves

Consider a string under a tension $F$ with its ends separated by a distance $L$. Fig. 2.1 depicts a complex wave on the string, which could be produced by plucking the string or drawing a bow across it. We will see that a complex wave such as this can be constructed from a sum of sinusoidal waves. Therefore, the focus of this experiment is on sinusoidal waves.

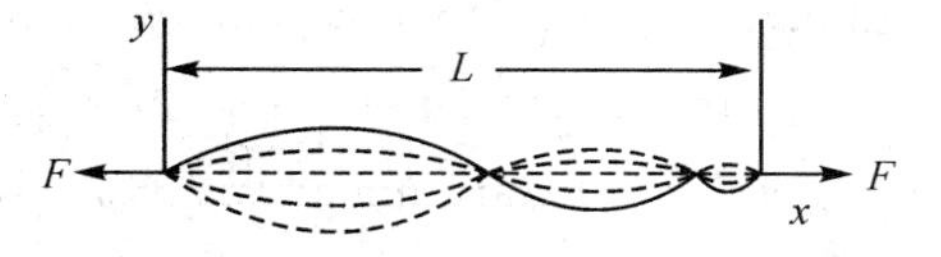


Fig. 2.1 A complex wave on the string.

Fig. 2.2 shows a wave traveling along the $x$-axis. The equation describing the motion of this wave is based on two observations. First, the shape of the wave does not change with time ($t$). Second, the position of the wave is determined by its speed in the $x$ direction. Based on these observations, we see that the vertical displacement of the wave ($y$) is a function of both $x$ and $t$.

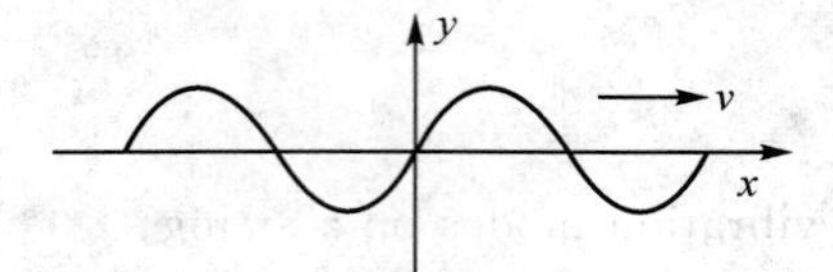


Fig. 2.2 A wave traveling along the $x$-axis.

Let $y(x, t=0)=f(x)$, where $f(x)$ represents the function that characterizes the shape of the wave. Then $y(x,t)=f(x-vt)$, where $v$ is the speed of the wave. Although this description holds true for all traveling waves, we will limit our discussion to sinusoidal waves.

The vertical displacement of the traveling sinusoidal wave shown in Fig. 2.2 can be expressed as

$$y(x,t)=A\sin\left[\left(\frac{2\pi}{\lambda}\right)(x-vt)\right] \tag{2.1}$$

where $A$ is the *amplitude* of the wave (i.e., its *maximum* displacement from equilibrium) and $\lambda$ is the *wavelength* (i.e., the distance between two points on the wave which behave identically).

Expanding the term inside the brackets gives

$$y(x,t)=A\sin\left(\frac{2\pi x}{\lambda}-\frac{2\pi vt}{\lambda}\right) \tag{2.2}$$

By substituting $\lambda=vT$, $k=2\pi/\lambda$, and $\omega=2\pi/T$ (where $T$ is the *period*, $k$ is the *angular wavenumber*, and $\omega$ is the *angular frequency*), we obtain

$$y(x,t)=A\sin(kx-\omega t) \tag{2.3}$$

As $t$ increases, the argument of the sine function $(kx-\omega t)$ decreases. In order to obtain the same value of $y$ at a later time, $x$ must also increase, which implies that this wave travels to the right. Conversely, the argument $(kx+\omega t)$ represents a wave traveling to the left. When the right-traveling wave of Fig. 2.3 reaches a fixed end of the string, it will be reflected in the opposite direction.

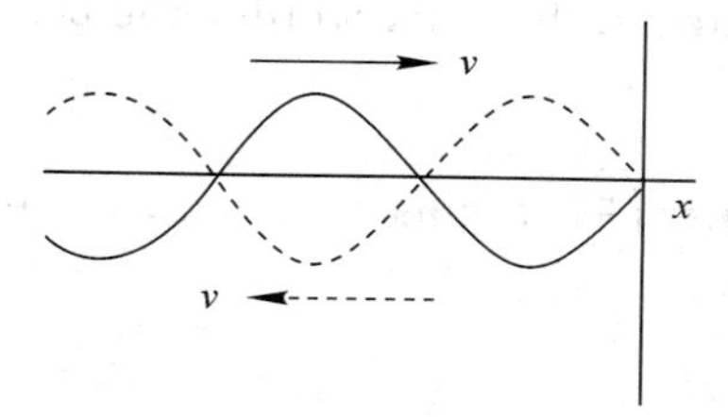


Fig. 2.3 The reflection of waves.

The right-moving incident wave, $y_1$, generates a left-moving reflected wave, $y_2$, with the same amplitude:

$$y_1(x,t)=A\sin(kx-\omega t) \tag{2.4}$$

$$y_2(x,t)=A\sin(kx+\omega t) \tag{2.5}$$

The resultant wave, $y_3$, which is the sum of the individual waves, is given by

$$y_3(x,t)=y_1(x,t)+y_2(x,t)=A\sin(kx-\omega t)+A\sin(kx+\omega t) \tag{2.6}$$

We can rewrite Eq. (2.6) by using the trigonometric identity:

$$A\sin(\alpha)+A\sin(\beta)=2A\sin\left(\frac{\alpha+\beta}{2}\right)\cos\left(\frac{\alpha-\beta}{2}\right) \tag{2.7}$$

$$y_3(x,t)=2A\sin(kx)\cos(\omega t) \tag{2.8}$$

Note that the $x$ and $t$ terms are separated such that the resultant wave is no longer traveling. Eq. (2.8) shows that all particles of the wave undergo simple harmonic motion in the $y$ direction with angular frequency $\omega$, although the maximum amplitude for a given value of $x$ is bounded by $|2A\sin(kx)|$. If we fix the two ends of the string and adjust the frequency so that an integral number of half waves fit into its length, then this *standing wave* is said to be in *resonance*.

The fixed ends impose a *boundary condition* on the string; its amplitude at the ends must be zero at all times. Thus, we can say that at $x=0$ and $x=L$ (where $L$ is the length of the string),

$$y_3(x=0,t)=y_3(x=L,t)=0 \tag{2.9}$$

$$2A\sin(k\cdot 0)\cos(\omega t)=2A\sin(kL)\cos(\omega t)=0 \tag{2.10}$$

or

$$\sin(kL)=0 \tag{2.11}$$

Eq. (2.11) is a boundary condition which restricts the string to certain modes of vibration. This equation is satisfied only when $kL=n\pi$, where $n$ is the *index* of vibration and is equal to any positive integer. In other words, the possible values of $k$ and $\lambda$ for any given $L$ are

$$kL=\left(\frac{2\pi}{\lambda}\right)L=n\pi\ (n=1,\ 2,\ 3,\ \cdots) \tag{2.12}$$

or

$$\lambda=\frac{2L}{n} \tag{2.13}$$

Fig. 2.4 shows the mode with index $n=4$: the fourth *harmonic*. The positions at which the vibration is small or zero are called *nodes*, while the positions where the vibration is largest are called *antinodes*. The number of antinodes is equal to the index of vibration and to the ordinal rank of the harmonic (fourth, in the case above). Fig. 2.5 shows several other modes of vibration.

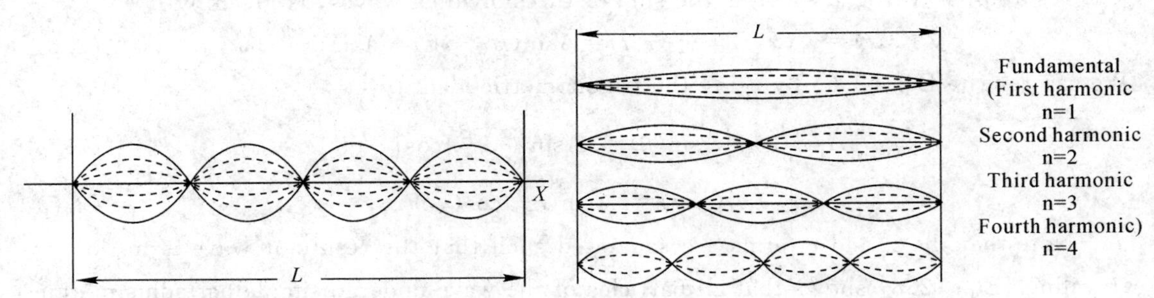


Fig. 2.4 The fourth *harmonic*.

Fig. 2.5 Several other modes of vibration.

Note that wavelength$=2\times$(string length between supports)$/n$, where $n$ is mode number (or number of antinodes).

2. Properties of standing waves

The wave speed ($v$) depends on two quantities—frequency ($f$) and wavelength ($\lambda$) which are related by

$$v=f\lambda \tag{2.14}$$

A wave is created by exciting a stretched string; therefore, the speed of the wave also depends on the tension ($F$) and the mass per unit length of the string ($\mu$). Physics texts give us the derivation of the wave speed:

$$v=\left(\frac{F}{\mu}\right)^{1/2} \tag{2.15}$$

By combining Eqs. (2.13), (2.14) and (2.15), we can express the frequency as

$$f=\frac{v}{\lambda}=\left(\frac{n}{2L}\right)v=\left(\frac{n}{2L}\right)\left(\frac{F}{\mu}\right)^{1/2} \quad (n=1, 2, 3, \cdots) \tag{2.16}$$

When a wave travels from one medium to another, some of its properties change (e. g. , speed and wavelength), but its frequency remains fixed. For example, consider the point where two strings of different densities are joined. If the two strings have different frequencies, then the two sides of the point would oscillate at their own frequencies, and the point would no longer be a "joint". Mathematically speaking, the function describing the string would not be continuous at the "joint". The constancy of the frequency allows us to determine the wave speed and wavelength in a different medium, if the frequency in that medium is known.

## Procedure

Set up the equipment. Adjust the vibrator clamp on the side to position it firmly in the vertical orientation. Run one end of the string from a vertical bar past the vibrator and over the pulley. The vibrator is connected to the string with an alligator clip. Attach a mass hanger to the other end of the string. Throughout this experiment, you will be changing the tension (by using different weights). To measure vibration amplitudes, it is helpful to have a meter stick clamped vertically near the string.

### Part 1

In this section, we will keep the tension and density of the string constant to find experimentally the relationship between frequency and number of antinodes.

1. Adjust the frequency until you obtain a nice standing wave with two antinodes ($n=2$). Record this frequency in the "Data" section.

2. Obtain and record the frequencies for consecutive $n$ values. Take at least six measurements, starting with the fundamental mode.

3. Calculate and record the wavelength, Eq. (2. 13), and wave speed, Eq. (2. 14), corresponding to each $n$.

4. Plot a graph of frequency as a function of $n$. What is the relationship between the two variables?

### Part 2

In this section, we will keep $n$ constant and change the weights to find the relationship between frequency and tension.

1. Choose one of the three strings. If you like to see data that agree well with theory, choose the finest string. If you would rather see more interesting data, for which you might need to explain the discrepancy, choose the most massive string. Measure and record the linear mass density ($\mu = M/L$) of the string by obtaining its total mass $M$ and total length $L$. Use the digital scale to weigh the string. Keep all units in the SI system (kilograms and meters).

2. Using the 50 g mass hanger, measure and record the frequency for the $n=2$ mode. (Note: You may choose any integer for $n$, but remember to keep $n$ constant throughout the rest of this section.)

3. Add masses in increments of 50 g, and adjust the frequency so that the same number of nodes is obtained. Take and record measurements for at least six different tensions.

4. The wave speed should be related to the tension $F$ and linear mass density $\mu$ by $v = (F/\mu)^{1/2}$. Calculate and record the wave speed in each case using Eq. (2.14), and plot $v^2$ as a function of $F/\mu$. (You have calculated $v^2$ from the frequency and wavelength; these are the $y$-axis values. You have calculated $F/\mu$ from the measured tension and linear mass density; these are the $x$-axis values. Be sure to convert the tension into units of Newtons.) You now have the experimental points.

5. Now plot the "theoretical" line $v^2 = F/\mu$. This is a straight line at 450 on your graph, if you used the same scale on both axes. Do your experimental and theoretical results agree well? If not, what might be the reasons?

**Part 3**

In this section, we will determine the relationship between frequency and the density of a medium through which a wave propagates.

1. Measure the linear mass densities ($\mu = M/L$) of the two other strings as described above.

2. Keep the tension and mode number constant at, say, 100 g and $n=2$, measure and record the frequencies for the three strings.

3. Calculate and record the "experimental" wave speed from the frequency and wavelength for each string density.

4. Calculate and record the "theoretical" wave speed for each string density from $v = (F/\mu)^{1/2}$, and compare these speeds with the experimental values.

## Observations and Results

**Table 2.1 Data records.**

The linear mass densities $\mu=$ ________ g/cm, $g=979.44$ cm/s$^2$

| Mass $m$/g | $\sqrt{m/g}$ | Mode $n$ | Length of string $L$/m | Wavelength $\lambda$/m | Wave speed $v$/(m/s) | Frequency $f$/Hz |
|---|---|---|---|---|---|---|
| 25 | | 6 | | | | |
| 75 | | 5 | | | | |
| 125 | | 4 | | | | |
| 200 | | 3 | | | | |
| 300 | | 2 | | | | |

## Calculation

1. Plot the graph of the frequency as a function of $n$ using one sheet of graph paper at the end of this workbook.

2. Plot the experiment graph of $v^2$ as a function of $F/\mu$ using one sheet of graph paper at the end of this workbook.

3. Plot the theoretical graph of $v^2$ as a function of $F/\mu$ using the same sheet of graph paper.

# Lab 3 Physical Pendulum

## Purpose

1. To measure the acceleration due to gravity by means of a physical pendulum;

2. To study the dependence of the period of a physical pendulum on other pendulum parameters.

## Apparatus

Pendulum consisting of a rod; photogate.

## Theory

The simple pendulum may be defined as a point mass attached to a massless unstretchable string, which is attached to a rigid support (See Fig. 3.1). Such a pendulum is, of course, only an idealization, but attaching a small dense sphere to the end of a long light string can make good approximation.

Any pendulum that is not a simple pendulum is, by definition, a physical pendulum or compound pendulum. A physical pendulum may be constructed by supporting a rigid body, symmetrical or otherwise, by some sort of pivot. The pivot may be located at any point except the center of mass (See Fig. 3.2).

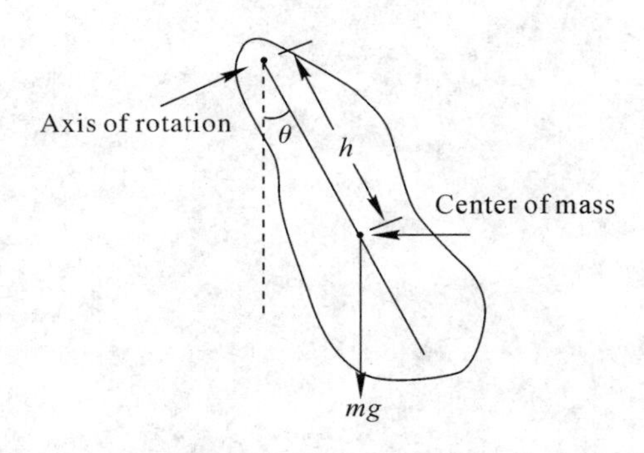


Fig. 3.1 Rigid body rotating about a fixed axis.

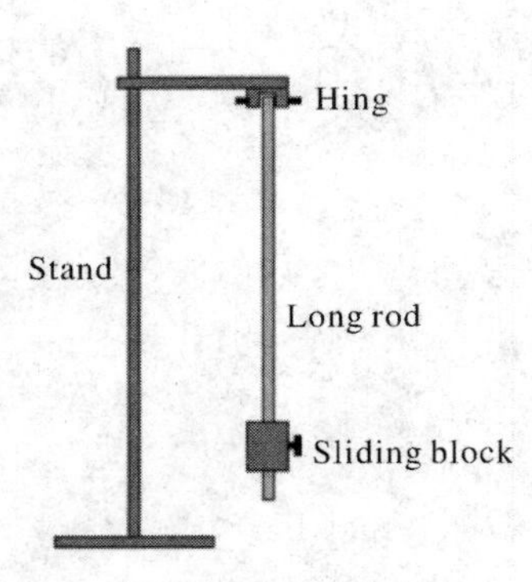


Fig. 3.2 Physical pendulum setup.

A physical pendulum is not a point mass on a string, but an object of any shape suspended from an axis through it. As it swings, it rotates about the axis. Thus, its motion is a non-uniform rotation. The torque ($\Gamma$) that causes this rotation is produced by the force of gravity acting at the center of mass.

Thus,

$$I\alpha=\Gamma=-mgh\sin\theta\approx-mgh\theta \tag{3.1}$$

where $I$, $\alpha$, and $h$ are moment of inertia about the axis, angular acceleration, and distance of center mass from the axis of rotation, respectively. For small angle oscillation, it can be written as

$$I\alpha\approx-mgh\theta \tag{3.2}$$

The angular frequency of the oscillation ($\omega$) is given by

$$\omega=\sqrt{\frac{mgh}{I}} \tag{3.3}$$

And the corresponding time period ($T$) is given by

$$T=2\pi\sqrt{\frac{I}{mgh}} \tag{3.4}$$

where $m$ is mass of the body, $h$ is the distance between center of mass (CM) and the axis of rotation, and $I$ is the moment of inertia (MI) about the axis of rotation given by (from parallel axis theorem).

$$I=I_0+md^2 \tag{3.5}$$

where $I_0$ is the moment of inertia about parallel axis through center of mass. If $k$ is the radius of gyration (i. e. , $I_0=mk^2$). Then from Eqs. (3.4) and (3.5)

$$T^2 d=\frac{4\pi^2}{g}(k^2+d^2) \tag{3.6}$$

By recording the period of oscillations $T$ as a function $d$ we can determine the values of gravitational acceleration $g$ as well as moment of inertia $I_0$ of the body. The plot of $T$ against $d$ shows a minimum time period at $d=k$, given by

$$T_{\min}=2\pi\left(\frac{2k}{g}\right)^{\frac{1}{2}} \tag{3.7}$$

## Procedure

In this experiment the rigid body consists of a rectangular mild steel bar with a series of holes drilled at regular interval to facilitate the suspension at various points along its length. The steel bar can be made to rest on screw type knife-edge fixed on the wall to ensure the oscillations on a vertical plane freely. The oscillations can be monitored accurately using a telescope. The radius of gyration for this bar is

$$k^2 = \frac{l^2 + b^2}{12} \tag{3.8}$$

where $l$ & $b$ are the length & breath of the bar, respectively.

1. Determine the CM by balancing the bar on a knife-edge. Measurement of $d$ is made from this point to the point of suspension for each hole.

2. Suspend the bar by means of knife-edge.

3. Place the pivot hole of the rod over the knife-edge. Set the pendulum into oscillation, with amplitude of about five degrees. Start the photogate as a pendulum comes to rest at one extreme position. Stop the stopwatch after 10 or 20 cycles. Record the elapsed time. Repeat for another 10 or 20 cycles.

4. Measure the time for 10 to 20 oscillations for different $d$(only on one side of CM). Repeat each observation several times.

5. Plot $T$ against $d$. Calculate $k$ and $g$ from this graph.

6. Plot $T^2 d$ Vs $d^2$. Using linear regression technique fit the data and determine $k$ and $g$ from it.

7. Calculate the average experimental value of $g$, and the percent error, using 980 cm/s$^2$ or 9.80 m/s$^2$ as the standard value.

## Observations and Results

**Table 3.1 Data records.**

| S. No. | Distance from CM of axis ($d$) in cm | No. of oscillations ($n$) | Time period for $n$ oscillations ($T_n$) | Time period ($T$) for 1 oscillation ($T_n/n$) | $T^2 d$ | $d^2$ |
|---|---|---|---|---|---|---|
| 1 | | | | | | |
| 2 | | | | | | |
| 3 | | | | | | |
| 4 | | | | | | |
| 5 | | | | | | |
| 6 | | | | | | |
| 7 | | | | | | |
| 8 | | | | | | |
| 9 | | | | | | |

## Calculation

1. $T_{min}$ from the graph =________, at $d$=________.

2. At $T_{min}$, $d=k=$________.

$$T_{min}=2\pi\left(\frac{2k}{g}\right)^{\frac{1}{2}}, \text{ so, } g=\_\_\_\_\_\_\_\_.$$

3. From the plot of $T^2d$ vs. $d^2$, find the slope and intercept from linear regression.

4. From the slope, $g$ can be calculated and from the intercept, $k$ can be calculated using the formula (3.8).

5. Compare with the value derived from the graph. Why are the two $k$ values different?

## Results

'$g$' value from $T$ vs. $d$ plot is ________.

'g' value from $T^2d$ vs. $d^2$ plot is ________.

Average value of '$g$'=________.

Using 9.80 $m/s^2$ as the standard value, the percent error of '$g$'=________.

# Lab 4　Moment of Inertia Measured by Three-Wire Pendulum

## Purpose

1. To master the principles and method of the moment of inertia measured by three-wire pendulum;

2. To master the correct method of measuring the length, the mass and the time;

3. To measure the moment of inertia of disk and ring round the axis of symmetry;

4. To test parallel axis theorem of the moment of inertia.

## Apparatus

Three-wire pendulum; bubble level; steel tape measure; stopwatch; tested ring; cylinder under test.

## Theory

Three-wire pendulum is symmetrical disc B hung under fixed and level small disc A by the same three wires (See Fig. 4.1). Disc B can swing back and forth round central axes $OO'$. The conversion between potential energy and the kinetic energy of disc B would happen in the swing process. The swing cycle is decided by moment of inertia of disc B and all objects on it. The moment of inertia of disc B and all objects on it may be determined according to swing cycle and the related geometry parameter.

Fig. 4.1　Three-wire pendulum.

The mass of disc B is $m_0$. When disc B swings with a small angle, the formula could be reduced when the friction and air resistance are neglected:

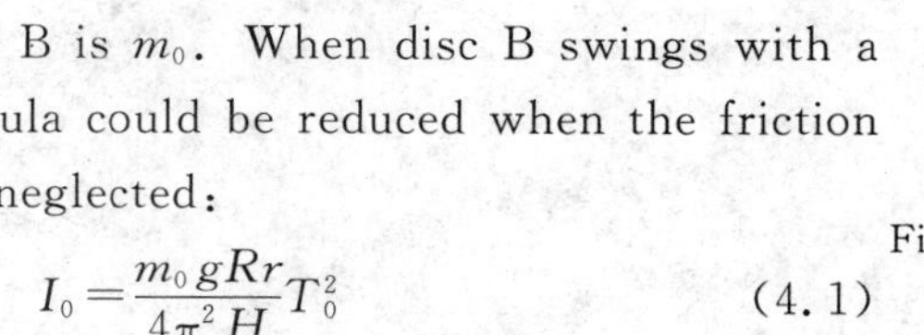

$$I_0 = \frac{m_0 gRr}{4\pi^2 H} T_0^2 \tag{4.1}$$

where $I_0$ is moment of inertia of disc B round its central axes $OO'$, $H$ is vertical distance between A and B, $l$ is length of hanging wire AB, $r$ and $R$ are distances from the joint to the centre of disc A and disc B, $T_0$ is the swing cycle of disc B.

If the ring whose mass is $m_1$ is set on disc B, spindle of the ring was kept according to central axes $OO'$, then $I_1$ is moment of inertia of the ring:

$$I_1=\frac{(m_1+m_0)gRr}{4\pi^2 H}T_1{}^2-I_0 \tag{4.2}$$

$T_1$ is the swing cycle of the ring and the disc B.

In theory, $I_r$ is moment of inertia of the ring round its central axes $OO'$:

$$I_r=\frac{1}{2}m\left[\left(\frac{d}{2}\right)^2+\left(\frac{D}{2}\right)^2\right]=\frac{1}{8}m(d^2+D^2) \tag{4.3}$$

where $m$ is the mass of the ring, $d$ and $D$ are internal and external diameters of the ring, respectively.

Moment of inertia of disc B round its central axes $OO'$ in theory can be expressed as $I_d$:

$$I_d=\frac{1}{8}m_0 D_0^2 \tag{4.4}$$

where $m_0$, $D_0$ are the mass and diameter of the disc B, respectively.

If two cylinders have the same shape, mass ($m_2$) and radius ($r_2$) and be put symmetrically on disc B, $d$ is distance from center of disc B, $I_c$ is the moment of inertia of a cylinder round the disc central shaft, then:

$$I_c=\frac{\frac{(2m_2+m_0)gRr}{4\pi^2 H}T_2^2-I_0}{2} \tag{4.5}$$

where $T_2$ is the swing cycle of both two cylinders and disc B. So parallel axis theorem may be confirmed from the experiment:

$$I_c=\frac{1}{2}m_2 r_2^2+m_2 d^2 \tag{4.6}$$

## Procedure

1. Adjust the three-wire pendulum and make disc A and disc B level.
2. Measure the parameters of the instrument, $l$, $R$, $r$, $H$, $d$, $D$, $r_1$ and $r_2$.
3. Measure $T_0$—the swing cycles of disc B, $T_1$—the swing cycle of the ring and the disc B, $T_2$—the swing cycle of both two cylinders and disc B. The swinging angle must be less than 5°.

4. Calculate the moment of inertia of the disc with formulas (4. 1) and (4. 4), respectively, calculate the relative errors.

5. Calculate the moment of inertia of the ring with formulas (4. 2) and (4. 3), respectively, calculate the relative errors.

6. Measure the moment of inertia of a cylinder round the disc central shaft, and test the parallel axis theorem of the moment of inertia using formulas (4. 5) and (4. 6).

7. The experimental data from of three-wire pendulum.

## Question

The rotation angle of the disc can not be too large, why?

# Lab 5 The Prism Spectrometer: Dispersion and the Index of Refraction

## Purpose

1. To calibrate a prism spectrometer with a standard (mercury) light source;
2. To measure the index of refraction of a glass prism for several wavelengths of light.

## Apparatus

Spectrometer; prism; mercury lamp.

## Theory

Fig. 5.1 shows a ray of light of a single wavelength, $\lambda$, incident upon an equilateral prism. Let the index of refraction of the glass at this particular wavelength be n. It is important to remember that the index of refraction depends on the wavelength. It is not a constant! We shall take the index of refraction of the surrounding air to be 1.0. Strictly speaking, this is not correct but our spectrometer is not precise enough to detect the difference. Let the apex angle of the prism be denoted by $A$. Its value can be determined accurately in the laboratory using the prism spectrometer.

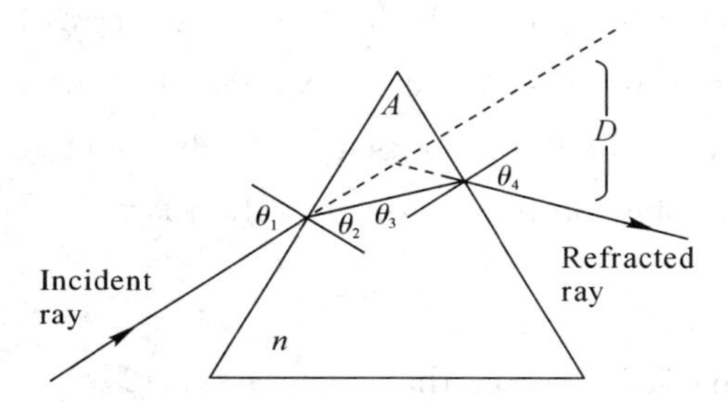


Fig. 5.1 Refraction of light by a prism.

We can apply Snell's Law to the ray of light on each surface. This leads to two equations,

$$\sin\theta_1 = n\sin\theta_2 \quad (5.1a)$$

$$n\sin\theta_3 = \sin\theta_4 \quad (5.1b)$$

The angles $\theta_2$ and $\theta_3$ are not independent, being related by the equation,

$$A = \theta_2 + \theta_3 \quad (5.2)$$

You are asked to prove this relation in your pre-lab. The angle $D$, known as the angle of deviation, is shown in Fig. 5.1. $D$ is the angle between the incident ray and the emergent ray. This angle is directly measurable in the laboratory. $D$ is the exterior angle to the small triangle formed by the dashed lines and the ray that passes through the prism. The exterior angle theorem from geometry says that the exterior angle to any triangle is equal to the sum of the two opposite interior angles of the triangle. These interior angles in our triangle are the angles $(\theta_1 - \theta_2)$ and $(\theta_4 - \theta_3)$. Applying the exterior angle theorem we get

$$D = (\theta_1 - \theta_2) + (\theta_4 - \theta_3) \quad (5.3)$$

Combining this equation with Equation (5.2), we have

$$D = \theta_1 + \theta_4 - A \quad (5.4)$$

For reasons to be discussed below, we would like to express $D$ in terms of the single angle $\theta_2$. This is accomplished by using Equations (5.1a) and (5.1b). Solving these equations for $\theta_1$ and $\theta_4$, respectively, we get

$$\theta_1 = \arcsin(n\sin\theta_2) \quad (5.5a)$$

$$\theta_4 = \arcsin(n\sin(A - \theta_2)) \quad (5.5b)$$

We have used the relation $\theta_3 = A - \theta_2$ from Equation (5.2) in Equation (5.5b). Finally, substituting these two equations into Equation (5.4) we get

$$D = \arcsin(n\sin\theta_2) + \arcsin(n\sin(A - \theta_2)) - A \quad (5.6)$$

There is a good reason for expressing the angle $D$ in terms of the single angle $\theta_2$. For a certain angle $\theta_2$, $D$ possesses a minimum value. To show this, one differentiates Equation (5.6) with respect to $\theta_2$ and sets the result equal to zero. This produces an equation for the angle $\theta_2$. The solution of this equation is,

$$\theta_2 = \frac{A}{2} \quad (5.7)$$

Now, from Equation (5.2) we see that when $\theta_2 = A/2$, $\theta_3 = A/2$ as well. This means that the angle $D$ has its minimum value $D_m$ when the incident ray is directed in such a way that the light passes through the prism symmetrically with respect to apex $A$. The angle $D_m$ is known as the *minimum angle of deviation for* the prism at the wavelength $\lambda$. It is this particular value of $D$ which we will measure in the laboratory. The experimental technique for finding $D_m$ is given in the procedure. Knowing $D_m$ for a particular wavelength, we can find the index of refraction of the material of the prism for that wavelength. Note that since $\theta_2=\theta_3$ when $D$ is at its minimum value, we see from Equations (5.1a) and (5.1b) (and from the symmetry of the problem) that $\theta_1=\theta_4$ as well. Then from Equation (5.4) we have

$$\theta_4=\frac{D_m+A}{2} \tag{5.8}$$

Substituting this into Equation(5.1b) gives

$$n=\frac{\sin((D_m+A)/2)}{\sin(A/2)} \tag{5.9}$$

The spectrometer (See Fig. 5.2)

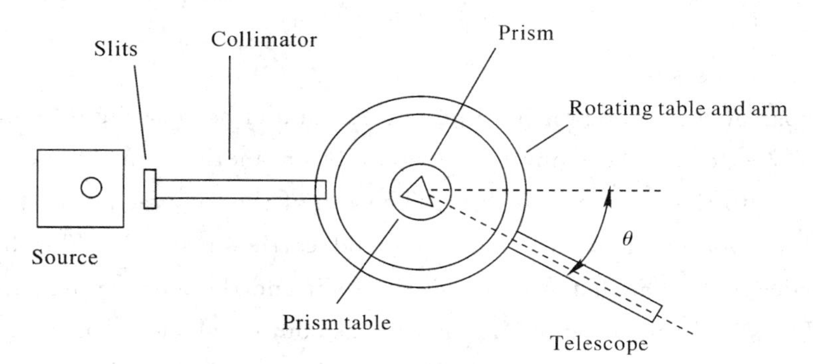


Fig. 5.2 Schematic diagram of the optical spectrometer using a prism.

1. Telescope

An astronomical (i.e. inverting) telescope with an achromatic objective and eyepiece is mounted on one arm of the spectrometer. The arm can be turned around a vertical axis passing through the center of the spectrometer. A graduated circular scale is attached to the telescope arm. This scale can be read by two verniers 180 degrees apart. The telescope stage can be clamped in any position by a screw and, in this position, a fine adjustment can be made by a tangent screw. Find these two adjustments and familiarize yourself with their operation.

The telescope tube may be slightly tilted with the help of two screws attached to the arm of the spectrometer. The telescope has an eyepiece which can be slid in and out to focus the crosshairs. The crosshairs can be rotated to make them horizontal and vertical.

(Try it.) Once the crosshairs are focused a rack and pinion arrangement is attached to the telescope tube to focus the telescope. Shown below in Fig. 5.3 is an Abbe telescope.

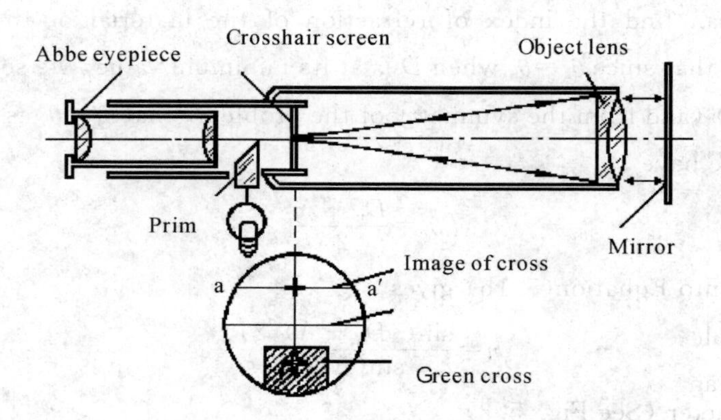


Fig. 5.3 Abbe telescope.

2. Collimator and slits

The purpose of the collimator is to produce a parallel beam of light, i. e. an object at infinity. It consists of a tube mounted horizontally on another arm of the spectrometer. This arm is fixed to the spectrometer base. The end of the tube facing the prism table has an achromatic converging lens and the other end carries a sliding tube attached to an adjustable vertical slit. The distance between the slit and the lens can be varied by a rack and pinion. The slit is composed of two sharp edges out of which one is kept fixed and the other can be moved by using the screw. There is also a slider to adjust the height of the slit. The entire "slit" can be rotated to either horizontal or vertical position.

3. Prism table

The prism table consists of two circular plates of the same radius separated by three springs. Each spring carries a leveling screw. Straight lines, parallel to a line that connects two of these screws, are engraved on the upper plate and the prism is placed on the table so that a reflecting surface is perpendicular or parallel to these lines. The vertical axis of rotation of the spectrometer passes through the center of the prism table. The height of the prism table can be adjusted with a clamping screw that fixes the table to the circular verniers. Thus the table can be moved around the vertical axis and its position (relative to the telescope arm scale) can be read by the verniers. Like the telescope arm it can also be fixed at any desired angle by a clamping screw and rotated very slowly using a tangent screw. Familiarize yourself with these adjustments.

4. The vernier scales

Use the vernier scale to read the angle to the nearest arc minute. (1arcmin=1′=1/60 degree.) Fig. 5.4 is an example.

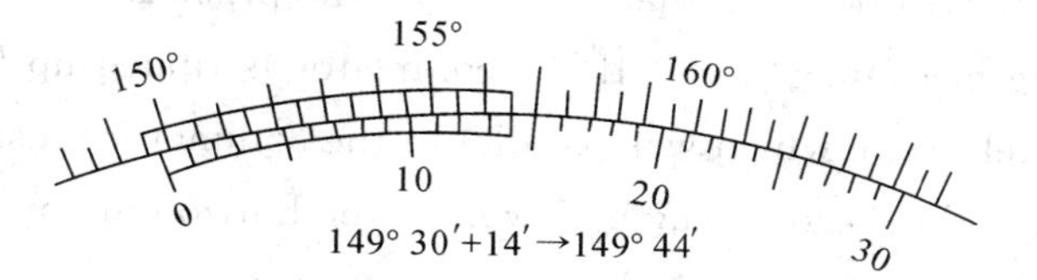


Fig. 5.4 A photograph of the vernier from the optical spectrometers.

In this example, the zero line of the vernier scale (the upper scale) is between 149.5° and 150°, so the angle is somewhere between 149°30′ and 150°. The vernier scale tells exactly where in between. Look along the vernier for the line that exactly lines up with the line below it. In this case, it is the 14′ line. So the angle is 149°44′, which we get by adding 14′ to 149°30′. Before using this angle in Equation (5.2), we must convert it to decimal degrees: 149+(44/60) degrees =149.73°.

## Procedure

1. Before measurements, we first need to adjust the spectrometer.

(a) Make the telescope and collimator parallel.

The collimator slit should now be turned horizontal (if not so). Adjust either the telescope or the collimator leveling screws to bring into coincidence the horizontal slit image and the intersection of the crosshairs. The telescope and collimator axes are now parallel, but not necessarily perpendicular to the mechanical instrument axis, as illustrated in Fig. 5.5.

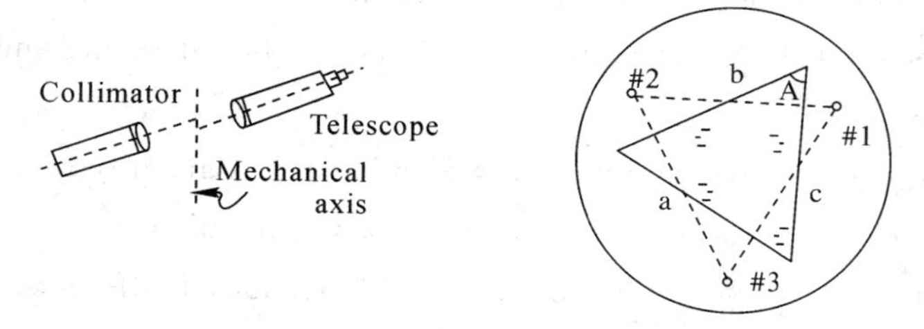


Fig. 5.5 Adjusting the telescope and the collimator.

(b) Make one prism face normal to the mechanical instrument axis.

Move the telescope so that it makes approximately a right angle with the collimator axis. Set one face of the prism to reflect light from the collimator into the telescope. Make the slit image coincide with the crosshairs by tilting the prism face. This face of the prism is now parallel to the mechanical axis. If the collimator is tipped up from the normal by, say, 2 degrees, the telescope which was parallel to the collimator must be down 2 degrees. The prism face bisects this vertical angle as well as the horizontal angle between collimator and telescope, and hence is normal to the instrument axis.

(c) Make the telescope axis normal to the adjusted prism face.

Without altering the level adjustment of the prism table, rotate the table so that a prism face is normal to the telescope. Using the Gaussian eyepiece, adjust the tilt of the telescope so that the telescope is now perpendicular to the prism face. The adjustment of the telescope is now completed.

(d) Again make the collimator and telescope axes parallel.

Adjust the collimator-axis to make it parallel to the axis of the telescope. Return to(b) above, with the telescope viewing the collimator at right angles via the prism. Adjust the collimator until coincidence between the slit and crosshairs is obtained. Now the collimator and telescope are both adjusted; they are parallel to each other, and the telescope is normal to the instrument axis.

(e) Make the collimator slit vertical.

The slit may be rotated to a vertical position with reasonable accuracy by eye or else by a reflection method after the prism is adjusted, provided the prism is a 60° one with 3 polished sides. The spectrometer is now in proper adjustment.

2. Prism Angle

When the prism is being used to view a spectrum, the light passes through only two of its faces. You must find the angle between these faces (the refracting angle). This angle is measured as follows:

Illuminate the slit with white light and widen it a bit. Carefully replace the prism table on the spectrometer. Locate the prism in such a position that the vertex of the refracting angle is toward the collimator. The prism should be placed off-center on the table, as shown, so that the reflected beams will not miss the telescope.

As shown in Fig. 5.6, the prism must remain in a fixed position for both readings.

Measure the angular position of the telescope for the two reflected beams (positions 1 and 2). Find the difference angle, $B$, between them. Derive the equation for prism angle $A$ from angle $B$. Measure the other two prism angles, and check the accuracy by how closely they add to 180°.

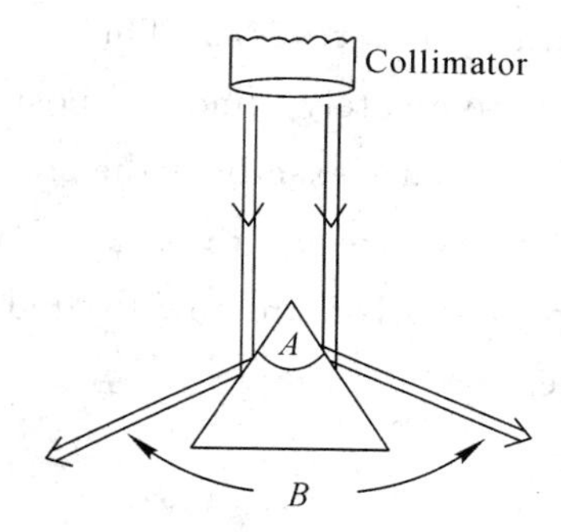


Fig. 5.6 Measuring the prism angle.

3. Minimum Deviation

The mercury green line has a well-established wavelength of 546.1 nm, so we will use it to measure the index of refraction of the prism glass at this wavelength by use of Equation (5.9).

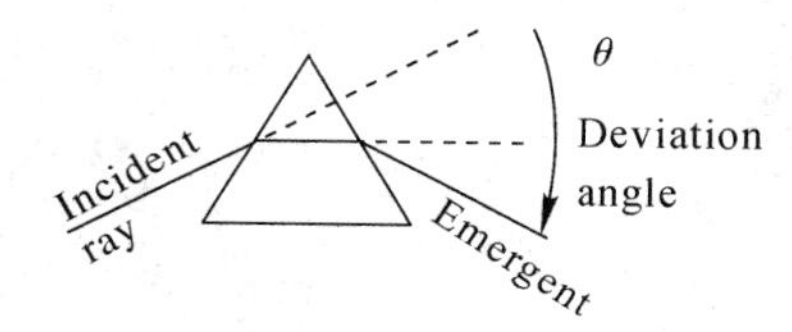


Fig. 5.7 Deviation angle of a prism.

Be sure the shaft collar on the prism table is tightened (gently) with the prism at the correct height. Loosen and remove the entire prism table. See straight through the telescope into the collimator at the slit, and read the position on scale for this "zero" setting. This "zero" value will then be used to correct subsequent readings to obtain true deviation angles (See Fig. 5.7).

Carefully replace the prism table, but leave it free to rotate. Locate the prism off-center on the table with its frosted side near the edge of the prism table. Carefully position the narrow neck of the mercury spectrum tube close to the slit of the collimator. Rotate the prism and telescope until you see the spectrum. Find the green line, and follow it with the telescope while you slowly rotate the prism table. If rotating the prism *increases* the angle of deviation, you are rotating it in the wrong direction. Rotate the prism in such a direction that the angle of deviation *decreases*. A point will be reached at which the spectral lines viewed in the telescope will slow their motion to a stop and reverse their motion, even though the prism table motion does not reverse. This is the minimum deviation angle. After having found this angle approximately, find the position of the telescope for which the reversal point of the green line occurs exactly at the crosshair intersection. Tighten the prism table clamping screw and do not change the prism position from now on. Read the spectrometer scale; the telescope and prism are now located to give the minimum deviation angle for the mercury green line.

## Questions

1. Calculate the index of refraction for wavelength of the mercury green line, and compare it with tabulated values. What kind of glass was used in *your* prism?

2. What absolute error in the index of refraction is caused by an error of one minute of arc in the deviation angle?

# Lab 6 The Transmission Diffraction Grating: Measuring the Wavelength of Light

## Purpose

1. To set up a diffraction-grating spectrometer for accurate measurements of wavelength;
2. To use the spectrometer to study the emission spectra of selected elements;
3. To obtain an estimate of the grating constant.

## Apparatus

Spectrometer; diffraction grating; mercury lamp and power supply; bubble level.

## Theory

A diffraction grating consists of a piece of metal or glass with a very large number of evenly spaced parallel lines or grooves. This gives two types of gratings: reflection gratings and transmission gratings.

Reflection gratings are ruled on polished metal surfaces; light is reflected from the unruled areas, which act as a row of "slits". Transmission gratings are ruled on glass, and the unruled slit areas transmit incident light.

The transmission type is used in this experiment. Common laboratory gratings have 300 grooves per mm and 600 groves per mm, and are pressed plastic replicas mounted on glass. Glass originals are very expensive.

Diffraction consists of the "bending", or deviation, of waves around sharp edges or corners. The slits of a grating give rise to diffraction, and the diffracted light interferes so as to set up interference patterns.

As shown in Fig. 6. 1, complete constructive interference of the waves occurs when the phase or path difference is equal to one wavelength, and the first-order maximum occurs for:

$$d\sin\theta_1 = \lambda \tag{6.1}$$

where $d$ is the grating constant, or distance between the grating lines, $\theta_1$ is the diffraction angle, and $d\sin\theta_1$ is the path difference between adjacent rays. The grating constant is given by

$$d = 1/N \tag{6.2}$$

where $N$ is the number of lines or grooves per unit length (usually per millimeter or per inch) of the grating.

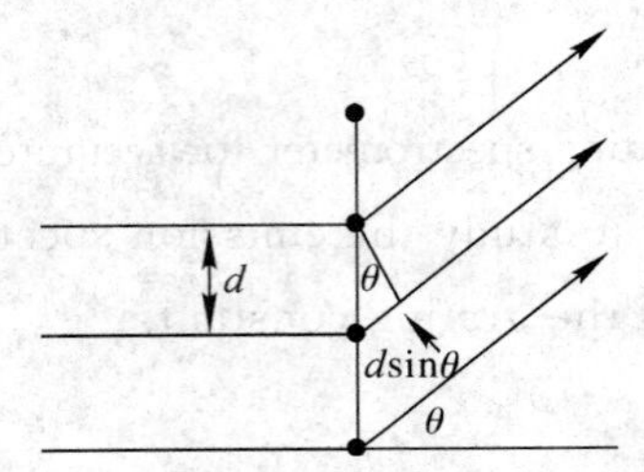


Fig. 6.1 Geometry determining the condition for diffraction from a multi-wire grating.

A diffraction grating is a multiple-slit device which produces spectral lines that are very sharp and widely spaced. Gratings can be made by ruling a large number of fine lines on a carefully prepared piece of glass or metal with a diamond point. In the case of glass, the transparent openings between the opaque lines constitute the parallel slit system. A grating will produce constructive interference for different wavelengths of light at an angle $\theta_n$ to the incident beam of light. The equation relating this diffraction angle to the wavelength and slit separation of the grating is

$$n\lambda = d\sin\theta_n, \quad n = 1, 2, 3, \cdots \tag{6.3}$$

where $\lambda$ is the wavelength of the diffracted light, $d$ is the grating space (distance between adjacent lines), and $n$ is the order of diffraction. That is, $n = 1$ for the first order diffraction, $n=2$ for the second order diffraction, and so on.

In this experiment you will be using a spectrometer and a diffraction grating to measure the various wavelengths of the spectral lines ofmercury lamp. A spectrometer is an instrument for producing and viewing the various lines (or wavelengths) of the spectrum. As shown in Fig. 6.2, it consists of a collimator $C$ to produce a parallel beam of light, a diffraction grating $G$ mounted on its table to disperse the light into its spectrum, and a telescope $T$ with which to examine the spectrum.

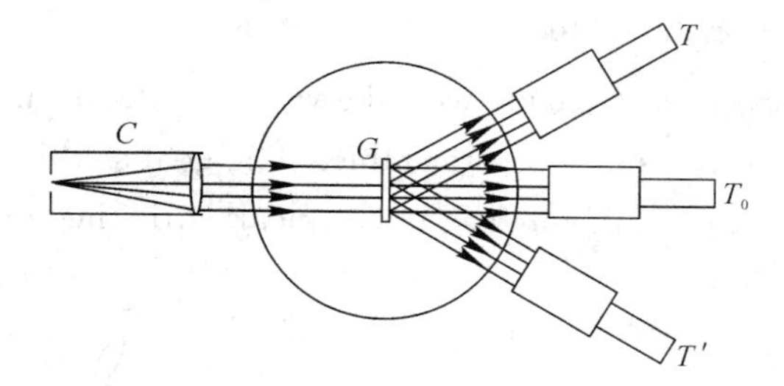


Fig. 6.2 Diffraction grating; telescope positions.

## Procedure

### Part 1 Adjusting Spectrometer

The main elements of the diffraction grating spectrometer are shown in Fig. 6.2. The derivation of the grating equation assumes that the incident light is in the form of plane waves. This is equivalent to assuming that the source of the light is infinitely far away (far-field approximation). In the spectrometer, this requirement is met by adjusting the collimator (Element $C$ in Fig. 6.2) such that the adjustable slit at the input end of the collimator is separated from a lens in the collimator by the focal length of the lens. This places the image of the slit at infinity. The plane waves emerging from the collimator pass through the diffraction grating (Element $G$) and interfere constructively along directions defined by the grating equation. For each wavelength, an image of the slit is formed, at the corresponding angle, using a telescope (Element $T$) focused at infinity.

Before the grating spectrometer is used, it is necessary to perform a number of adjustment steps as follows:

**Collimator Focus**

A simpler wayto focus the collimator is to carry the spectrometer (without a grating in place) to a window from which it is possible to view distant objects. By focusing the telescope on a VERY distant object or building, you will adjust it so that plane waves could focus to a sharp image. Without disturbing the telescope focus, return to the lab and view a discharge tube through the collimator slit, the collimator lens, and the telescope. Adjusting the collimator lens to give a sharp image of the slit through the telescope (which should still be focused at infinity) should ensure that image of the slit produced by the collimator lens is effective at infinity.

**Leveling of the Spectrometer Table**

Fig. 6.3 shows the spectrometer table with the vernier scale for measurement of angles and three screws for leveling of the spectrometer table. The goal at this step of adjustment is to ensure that the normal to the table surface is co-linear with the rotation axes of the table and the telescope mount.

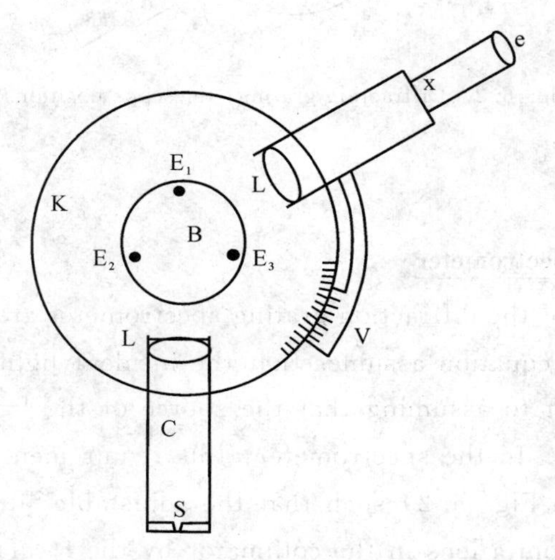


Fig. 6.3 The spectrometer table showing leveling screws and vernier scale.

Use a bubble indicator to level the baseplate, the collimator, and the telescope, and to perform the initial levelling adjustment on the spectrometer table, B.

Next, use the Hg discharge to fully illuminate the collimator slit. Rotate the telescope to the straight-through position so that the slit image is centered in its field of view. Swing the telescope in the direction for easiest viewing and clamp it at 90° to the straight-through position.

Place a prism on the spectrometer table so that either of its polished faces can be rotated into position to reflect light from the collimator into the telescope. Rotate the spectrometer table so that one of the reflected images is visible through the telescope, and note its vertical position in the field of view. Rotate the spectrometer table to bring the corresponding image from the other prism surface into view. If the two images do not appear at the same height in the field of view, then the prism table is not accurately perpendicular to the rotation axis of the instrument. Rotate the spectrometer table back and forth to view the two images in turn, and use the leveling screws to remove any difference between their heights. When the grating is positioned on the table, it may be necessary to adjust the position of the grating in its holder to ensure that spectral lines at large angles remain centered in the telescope.

**Alignment of the Diffraction Grating**

The grating should be accurately aligned perpendicular to the incoming light beam using the following procedure. Note the vernier reading with the undispersed image centered on the crosshair with the telescope in the straight-through position and the grating removed.

Rotate the telescope through 90° and clamp it in position. Mount the diffraction grating at the center of the spectrometer table. Rotate the grating so that it reflects the undispersed light toward the telescope. Rotate the table by precisely 45° so that the grating is normal to the collimator axis. Clamp the main vernier scale. Unclamp the telescope and rotate it by precisely 90° back to the straight through position ($T_0$ in Fig. 6. 2). Check that the image is again centered on the crosshairs and record the vernier reading as the main point of reference corresponding to $\theta_0 = \theta_0' = i = 0$.

Use the Hg green line to measure the diffraction angles $\theta_1$ and $\theta_1'$: these angles correspond to telescope positions $T$ and $T'$ in Fig. 6. 2. If the two angles differ by more than a few minutes of arc, the procedure should be repeated.

**Part 2 Measuring Wavelength**

Turn on the mercury lamp. The collimator has been adjusted so that the light coming through the slit comes out as parallel rays. The grating has also been adjusted so that its plane is perpendicular to the incident rays. Be careful not to move the grating table. If you do, do not touch anything else and re-align it by using the accepted value for wavelength for one of the spectral lines in the table below and the expected diffraction angle. You can do this by setting the telescope arm on the expected diffraction angle and rotating the table until the spectral line lines up with the crosshairs. To double-check that the grating is indeed perpendicular to the incoming beam, swing the telescope to the other side of the grating and make sure you obtain *the same reading of angle* for the same spectral line on this side.

**Using the spectrometer and Equation** (6. 3), **find the value of wavelength for three different spectral lines.** Similar to the previous lab where you calculated the index of refraction using the spectrometer, use the verniers to make measurements of the diffraction angle of spectral lines to the nearest minute of arc. Compare your values of the wavelengths with the accepted values in Fig. 6. 4 and calculate the percent error.

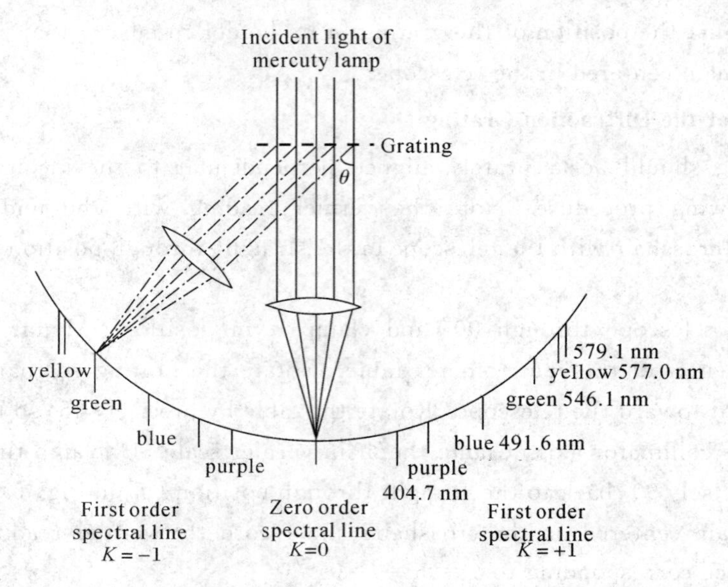


Fig. 6.4 Light spectrum of mercury lamp grating.

Also in each case determine the experimental error in the wavelength. Be careful when propagating error with angles! If you have an expression that multiplies something by, such as $\Delta C = \sin\theta \cdot \Delta\theta$, that had better be in radians. Any time you mess around with angles *outside of a trig function*, use *radians*! The data for this part of the experiment consists solely of a table. Be sure to include all relevant information and label it clearly.

**Part 3 Finding an Unknown Grating Spacing**

(1) Turn the telescope until the bright green mercury line of the first order appears on the crosshairs. Make two independent settings on this line and record your results on the data sheet.

(2) Repeat (1) with the first order on the other side of the central image.

(3) Repeat (1) and (2) for all other visible mercury lines, taking only one setting on each side of the central image.

(4) Repeat (1) and (2) for the second order bright green mercury line.

## Questions

1. You made two determinations of the wavelength of the bright green mercury line, one based on the first order interference and the other based on the second order interference. Which one do you think is more precise? Explain why.

2. How narrow should the slit be to just resolve the two spectral lines?

# Lab 7 Interference of Light

## Purpose

1. To illustrate the interference pattern produced from the air wedge and how it is formed;

2. To measure the thickness of the thin object using the interference fringes;

3. To explain the formation of Newton's rings;

4. To measure the wave length of the monochromatic light (Sodium).

## Apparatus

Travelling microscope; sodium lamp; glass plate; thin object, Newton's ring.

## Theory

When light is reflected from the two surfaces of a very thin film of varying thickness, an interference pattern is produced. Wherever the two reflected waves are in phase, bright areas appear. Whereever the difference in phase is one-half wavelength, dark areas are produced.

### Part 1 Air Wedge

A thin film having zero thickness at one end and *progressively* increasing to a particular thickness at the other end is called a wedge. A thin wedge of air film can be formed by two glasses sliding on each other at one edge and separated by a thin spacer at the opposite edge. The arrangement for observing interference of light in a wedge shaped film (shown in Fig. 7.1a). The wedge angle is usually very small and of the order of a degree. When a parallel beam of monochromatic light illuminates the wedge from above, the rays reflected from the two bounding surfaces of the film are not parallel and they appear to diverge from a point near the film. These rays interfere constructively or destructively producing alternate bright and dark fringes (See Fig. 7.1b).

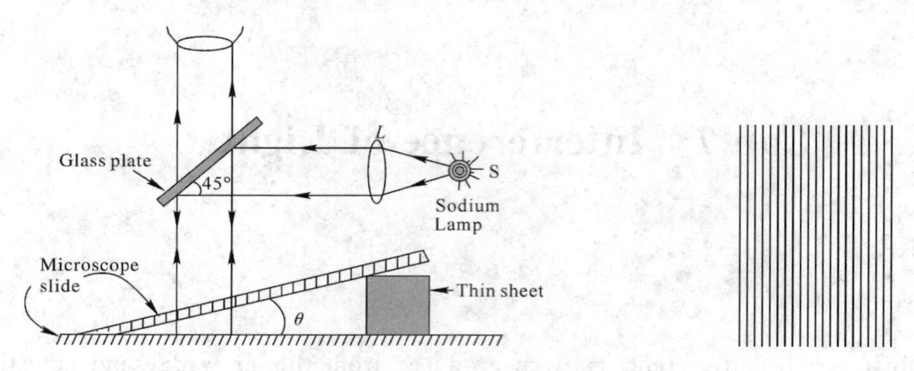


Fig. 7. 1a Interference of light in a wedge shaped film. Fig. 7. 1b Alternate bright and dark fringes.

When the light is incident on the wedge from above, it gets partly reflected from the glass-to-air boundary at the top of the air film. The other part of the light is transmitted through the air film and gets reflected at the air-to-glass boundary (as shown in Fig. 7. 2).

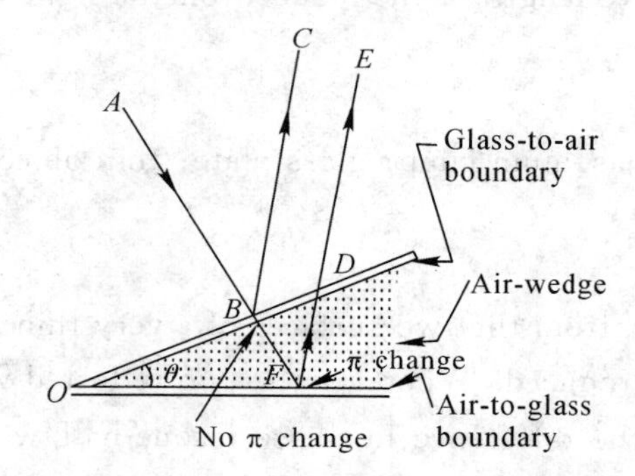


Fig. 7. 2 Reflection at the air-to-glass boundary.

The two rays *BC* and *DE* reflected from the top and the bottom of the air film have a varying path difference along the length of the film due to variation of the film thickness, because ray *DE* travels more distance than *BC*. Also, ray *DE* undergoes a phase change of half wave length ($\pi$ change) which occurs at the air-to-glass boundary due to reflection. The optical phase difference between the two rays *BC* and *DE* is given by

$$\Delta=2nt+\frac{\lambda}{2} \tag{7.1}$$

Minima occurs when the phase difference is an odd multiple of $\lambda/2$. The two waves arriving are 180° out of phase and give rise to destructive interference. Therefore, the condition for dark fringes, or destructive interference is:

$$\Delta=\left(m+\frac{1}{2}\right)\lambda \tag{7.2}$$

$$2nt=m\lambda$$

Because the film produced from air $n=1$:

$$2t=m\lambda \tag{7.3}$$

The thickness of the spacer used to form the wedge which shapes air film between the glass slides can be determined using a travelling microscope.

$$t=\frac{L\cdot\frac{\lambda}{2}}{d} \tag{7.4}$$

where $t$ is thickness of the spacer, $L$ is the length of the glass piece, $\lambda$ is the wave length of the used monochromatic light (sodium) in vacuum, and $d$ is the thickness of the fringe.

Steps:

1. Set the apparatus as shown in (Fig. 7.1a).
2. Fix the crosshair to one of the parallel fringes produced; take the readings of the vernier at one of the dark fringes ($d_0$).
3. Take the reading again after counting 20 dark fringes from the previous one ($d_{21}$).
4. Calculate $d$ using $d=(d_{21}-d_0)/20$.
5. Measure the length of the glass piece, starting from the edge of the thin spacer to the end of the plate($L$).
6. Calculate the thickness of the plate using Equation (7.4).

**Part 2 Newton's Rings**

Newton's rings is a phenomenon in which an interference pattern is created by the reflection of light between two surfaces — a spherical surface and an adjacent flat surface. It is named after Isaac Newton, who first studied them in 1717. When viewed with monochromatic, Newton's rings appear as a series of concentric, alternating bright and dark rings centered at the point of contact between the two surfaces. When viewed with white light, it forms a concentric-ring pattern of rainbow colors, because the different wavelengths of light interfere at different thicknesses of the air layer between the surfaces.

The bright rings are caused by constructive interference between the light rays reflected from both surfaces, while the dark rings are caused by destructive interference. Also, the outer rings are spaced more closely than the inner ones. Moving outwards from one dark ring to the next, for example, increases the path difference by the same amount, $\lambda$, corresponding to the same increase of thickness of the air layer, $\lambda/2$. Since the slope of the convex lens surface increases outwards, separation of the rings gets smaller for the outer rings. For surfaces that are not convex, the fringes will not be rings but will have other shapes.

Newton's rings are formed when a Plano-convex lens of large radius of curvature is placed on a plane glass sheet. The combination forms a thin circular air film of variable thickness in all directions around the point of contact of the lens and the glass plate at *O*. If monochromatic light is allowed to fall normally (Fig. 7.3a) on the lens using the 45° inclined glass plate, and the film is viewed in reflected light, interference fringes are observed in the form of a series with concentric rings (Fig. 7.3b).

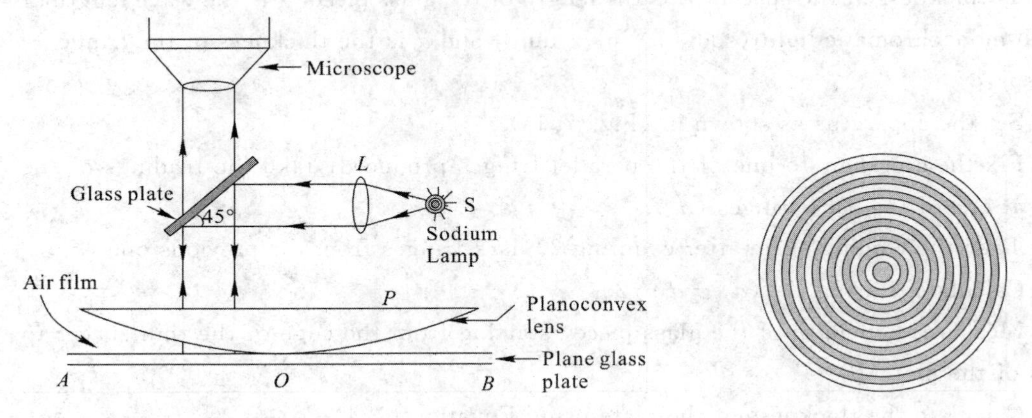


Fig. 7.3a Interference of light in a Plano-convex lens. Fig. 7.3b A series with concentric rings.

When the light is incident on the Plano-convex lens part of the light incident on the system is reflected from glass-to-air boundary (say at point *D*). The remainder of the light is transmitted through the air film. It is again reflected from the air-to-glass boundary (say from point *J*). See Fig. 7.4. The two rays (1 and 2) are reflected from the top and bottom of the air film interfere with each other to produce darkness and brightness.

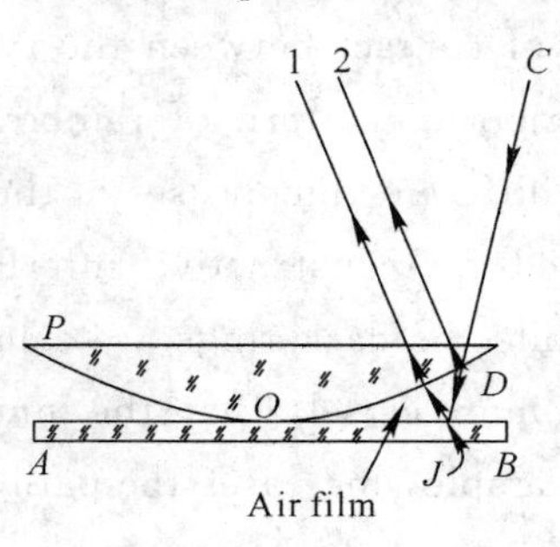


Fig. 7.4 Reflections from the top and bottom of the air film.

The condition for destructive interference is the same as obtained from the air wedge experiment:

$$\text{Constructive interference} \qquad 2t=(2m+1)\frac{\lambda}{2}$$

$$\text{Destructive interference} \qquad 2t=m\lambda$$

**Central dark spot**: At the point of contact of the lens with the glass plate the thickness of the air film is very small compared to the wavelength of light. Therefore, the path difference introduced between the interfering waves is zero. Consequently, the interfering waves at the centre are opposite in phase and interfere destructively. Thus a dark spot is produced.

**Circular fringes with equal thickness**: Each maximum or minimum is a locus of constant film thickness. Since the locus of points having the same thickness fall on a circle with its centre at the point of contact, the fringes are circular.

**Fringes are localized**: Though the system is illuminated with a parallel beam of light, the reflected rays are not parallel. They interfere nearer to the top surface of the air film and appear to diverge from there when viewed from the top. The fringes are seen near the upper surface of the film and hence are said to be localized in the film.

$$\text{Radii of the } m^{\text{th}} \text{ dark rings: } r_m=\sqrt{m\lambda R}$$

$$\text{Radii of the } m^{\text{th}} \text{ bright rings: } r_m=\sqrt{(2m+1)R\frac{\lambda}{2}}$$

The radius of a dark ring is proportional to the radius of curvature of the lens by the relation $r_m \propto \sqrt{R}$. Rings get closer as the order increases ($m$ increases) since the diameter does not increase in the same proportion.

*In transmitted light the ring system is exactly complementary to the reflected ring system so that the centre spot is bright. Under white light we get colored fringes.*

The wavelength of monochromatic light can be determined as

$$\lambda=\frac{D_{m+p}^2-D_m^2}{4pR} \tag{7.5}$$

where $D_{m+p}$ is the diameter of the $(m+p)^{\text{th}}$ dark ring and $D_m$ is the diameter of the $m^{\text{th}}$ dark ring.

## Procedure

1. Turn on the sodium lamp and adjust the apparatus so we have a parallel light falling

in the lens and the rings are seen clearly in the eyepiece of the travelling microscope.

2. Fix the crosswire on $20^{th}$ ring either from right or left of the centre dark ring and take the readings.

3. Move the crosswire and take the reading of $20^{th}$, $19^{th}$…$11^{th}$ rings.

4. You have to take the reading of rings on either side of the centre dark ring.

5. Enter the readings in the tabular column.

6. Calculate the radius of curvature of a lens by using the given Equation (7.5).

## Observations

The readings should be obtained as follows:

**Table 7.1 Data records.**

| Order of ring/m | Microscopic reading/mm | | Diameter $D_m$/mm | $D_m^2$/mm$^2$ | $D_{m+5}^2-D_m^2$ /mm$^2$ |
|---|---|---|---|---|---|
| | Left | Right | | | |
| 20 | | | | | |
| 19 | | | | | |
| 18 | | | | | |
| 17 | | | | | |
| 16 | | | | | |
| 15 | | | | | Mean value of $D_{m+5}^2-D_m^2$ |
| 14 | | | | | |
| 13 | | | | | |
| 12 | | | | | |
| 11 | | | | | |

## Calculation

The radius of curvature of the lens: $R=\bar{R}\pm\Delta\bar{R}$(m)= ______ ± ______ (m).

# Lab 8 Basic Measurement

## Purpose

1. To measure the length of an object, inter and outer diameter, depth and capacity of a cup;

2. To measure the diameter of a metal wire, and the thickness of a piece of object;

3. To measure the width of a single-slit and double-slit.

## Apparatus

Vernier caliper; screw micrometer; microscope; hollow metal cuboids with single-slit and a cylinder.

## Theory

### Optical Microscope

1. The components of microscope

There are two optical systems in a compound microscope: eyepiece lenses and objective lenses:

**Eyepiece or Ocular** is what you look through at the top of the microscope. Typically, standard eyepieces have a magnifying power of 10x. Optional eyepieces of varying powers are available, typically from 5x-30x.

**Eyepiece Tube** holds the eyepieces in place above the objective lens.

Objective lenses are the primary optical lenses on a microscope. They range from 4x-100x and typically, include three, four or five on lens on most microscopes. Objectives can be forward-or rear-facing.

**Coarse and Fine Focus knobs** are used to focus the microscope. Increasingly, they are coaxial knobs—that is to say, they are built on the same axis with the fine focus knob on the outside. Coaxial focus knobs are more convenient since the viewer does not have to grope for a different knob.

**Stage** is where the specimen to be viewed is placed. A mechanical stage is used when working at higher magnifications where delicate movements of the specimen slide are required. Stage clips are used when there is no mechanical stage. The viewer is required to move the slide manually to view different sections of the specimen.

**Aperture** is the hole in the stage through which the base (transmitted) light reaches the stage; Tube connects the eyepiece to the objective lenses.

**Illuminator**: A steady light source used in place of a mirror. The microscope has a mirror; it is used to reflect light from an external light source up through the bottom of the stage.

2. The working principle of microscope

The optical microscope magnifies an object in two steps. In both steps optical systems acting like converging lenses are used, as shown in Fig. 8.1.

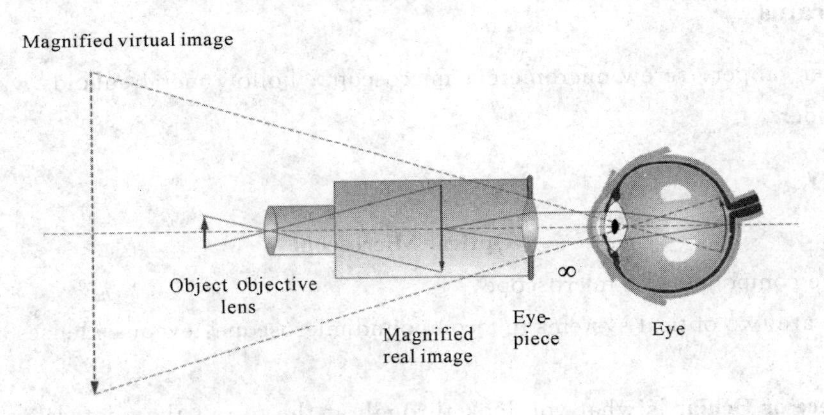


Fig. 8.1 Compound microscope schematic.

The first step is to place the object between the single and double focal point. The result is a magnified, real image. This microscope lens (in reality an optical system consisting of several lenses) is called the objective.

Then a second lens is used to pick up this image exactly in its front focal point. As a result, we generate a beam of parallel rays, but not a real image. This optical element is called eyepiece. The human eye is able to handle this parallel beam and generates an image onto its retina.

Finally, this is what one can expect from a microscope: objects can be observed on a magnified scale with details undetectable by the naked eye.

3. How to use a microscope

① When moving your microscope, always carry it with both hands. Grasp the arm with one hand and place the other hand under the base for support.

② Your microscope slide should be prepared with a coverslip or cover glass over the specimen. This will help protect the objective lenses if they touch the slide. Place the microscope slide on the stage and fasten it with the stage clips. You can push down on the back end of the stage clip to open it.

③ Look at the objective lens and the stage from the side and turn the coarse focus knob so that theobjectives lens moves downward (or the stage, if it moves, goes upward). Move it as far as it will go ***without touching the slide*** !

④ Now, look through the eyepiece and adjust the illuminator (or mirror) and diaphragm for the greatest amount of light.

⑤ Slowly turn the coarse adjustment so that the objective lens goes *up* (away from the slide). Continue until the image comes into focus. Use the fine adjustment, if available, for fine focusing. If you have a microscope with a moving stage, then turn the coarse knob so the stage moves downward or away from the objective lens.

⑥ Move the microscope slide around so that the image is in the center of the field of view and readjust the mirror, illuminator or diaphragm for the clearest image.

⑦ The proper way to use a monocular microscope is to look through the eyepiece with one eye and keep the other eye open (this helps avoid eye strain). If you have to close one eye when looking into the microscope, it is okay. Remember, everything is upside down and backwards. When you move the slide to the right, the image goes to the left!

⑧ Do not touch the glass part of the lenses with your fingers. Use only special lens paper to clean the lenses.

### Vernier Caliper

The construction of a vernier caliper (0. 05 mm in accuracy for example) is shown in Fig. 8. 2. F is the main ruler and G, which is attached to F, is the shifting ruler and is able to slide. Pincers A and B are used for measuring the outer diameter of a cup; and pincers C and D are for measuring of the inner diameter of a cup. T is used for measuring the depth of a container. Nut S can fasten the shifting ruler F on the main ruler. Length of 20 ticks on G is equal to 39 mm, as labeled on F, which means the length per tick on G is equal to 1. 95 mm. The difference between a tick on F and two ticks (2 mm) on G is 0. 05 mm, and we utilize this feature for a more accurate measurement.

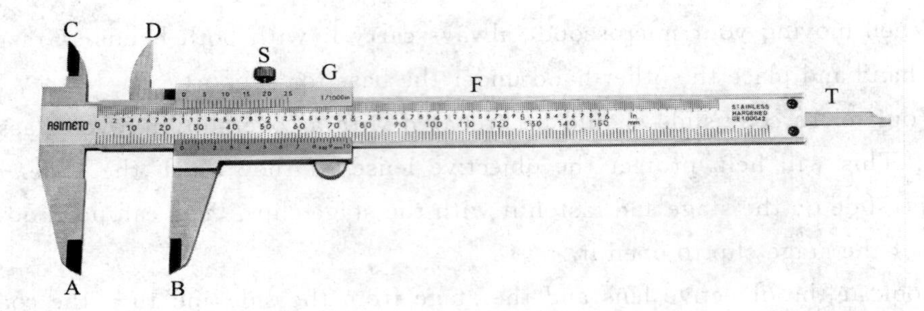


Fig. 8.2 The construction of a vernier caliper.

Take Fig. 8.3 for example. Pincers A and B grip an object and the 0th tick of G falls between 16 and 17 mm of the main ruler. We then have to find out the tick on G which matches (or being the closest) to one of the ticks on F. As shown in Fig. 8.3, 1.5 (3rd tick) on G perfectly matches a tick on F, so the length is measured as

$$(16+3\times0.05)\pm0.025=16.15\pm0.025 \text{ mm}$$

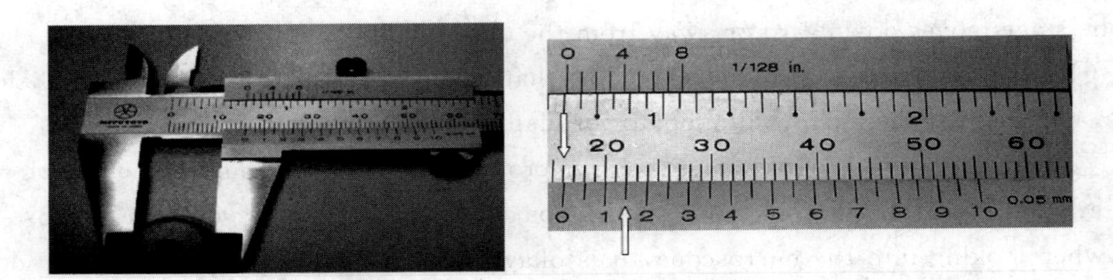


Fig. 8.3 The example of vernier caliper.

## Screw Micrometer

Screw micrometer can get a more accurate measurement of the thickness of an object. The structure of a screw micrometer is shown in Fig. 8.4. C is a scale holder. A is a fixed end, and P is connected with K′ and installed through F and K. On the periphery of H (the left-end of K) are 50 ticks which function as a sub-meter. P shifts 1 mm when K is screwed two rounds. Therefore, P shifts 0.01 mm when K is screwed by 1 tick. When measuring, put the object between A and P, screw K′ to clamp the item between A and P until 3 clicks. Find out the ticks on S (main meter) of which the edge of H falls between, and then read the ticks on the sub-meter H. Eventually, we can get the accurate measurement.

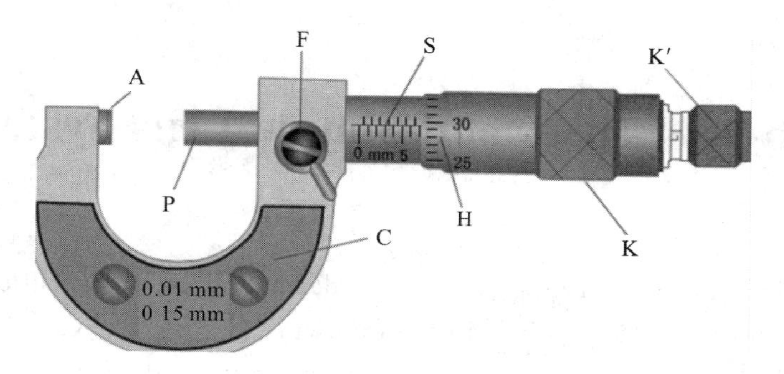


Fig. 8.4  The construction of a screw micrometer.

## Procedure

1. Use the metric scale of the vernier caliper to make three measurements of the diameter of a coin and record these measurements. Practice on other objects until you are sure you know how to read the instrument.

2. Examine the micrometer caliper and watch how the scale reading changes as you carefully close the jaws. Be sure you do not force the screw. If your instrument should have correction record it as + or −, depending on whether it is to be added to or subtracted from actual instrument reading.

3. Measure the thickness of a sheet of paper in your book with the micrometer caliper. Make three trials by using different sheets, record and determine the average. Now measure the total thickness of about all sheets in your textbook with the vernier caliper, three trials, then determine the average thickness of a single sheet in mm and record.

4. Measure the length of the metal plate with the ruler and the thickness of the metal plate with a micrometer caliper. Measure the length of the slit and the diameter of the small circular hole in the metal plate with a Vernier caliper. Measure the width of the slit with microscope, three trials of each record, and determine the average. Then estimate the uncertainty of measurement for each.

5. Compute the volume of the metal plate and the uncertainty.

# Lab 9 Lenses and Simple Lens Systems

Lenses produce different types of images, depending on the position of the object relative to the lens and its focal point. In the following discussion, we use the descriptions of real or virtual, erect or inverted, and reduced or magnified to describe these images. The fundamental distinction between real and virtual images is that light rays converge to form a real image, while light rays diverge to form a virtual image. Real images may be projected onto a screen while virtual images must be viewed through a lens system such as the eye.

## Purpose

1. To verify the thin lens equation and to construct simple optical systems using various combinations of lenses;
2. To determine the focal length of the single and also of a combination lens.

## Apparatus

Optical bench; converging lens; diverging lens; lens holder; light source; object screen; image screen; plane mirror.

## Theory

A lens is an image forming device. One of the simplest ways to describe the passage of light through a lens and the formation of an image is by ray optics. Ray optics assumes a beam of light is made up of many beams of light obeying Snell's Law at each boundary.

The *principal axis* of a lens is a line drawn through the center of the lens perpendicular to the face of the lens. The *principal focus* is a point on the principal axis through which incident rays parallel to the principal axis pass, or appear to pass, after refraction by the lens.

The *focal length* of a lens is the distance from the optical center of the lens to the principal focus; the reciprocal of the focal length in meters is called the *power of a lens* and is expressed in diopters. If two thin lenses whose focal lengths are $f_1$ and $f_2$, respectively, are placed in contact to be used as a single lens, the focal length, $F$, of the combination is given by the equation

$$\frac{1}{F}=\frac{1}{f_1}+\frac{1}{f_2} \tag{9.1}$$

In this experiment, we use a simple "optical bench," a magnetized track upon which various combinations of objects and lenses can be mounted. The distance between a lens and object, $d_o$ is called the object distance. The distance between a lens and image, $d_i$ is called the image distance. Denoting the focal length of a lens by, $f$, we may use the optical bench to verify the thin-lens equation for a given lens, image distance, and object distance:

$$\frac{1}{d_o}+\frac{1}{d_i}=\frac{1}{f} \tag{9.2}$$

The other way to determine the focal length of the lens is Bessel's method described as follows.

A lens is movedalong the optical axis between a fixed light source and a fixed screen. The light source and image are separated by a distance $L$ that is more than four times the focal length of the lens. Two positions of the lens are found for which an image is in focus on the screen, magnified in one case and reduced in the other. If two lens positions differ by distance $d$, then the focus length of the lens is given by

$$f=\frac{L^2-d^2}{4L} \tag{9.3}$$

In the diagram above, the points labeled $f$ are the focal points of the lens. *The lateral magnification*, $m$, of the image is defined as $m=|h_i/h_o|$. One can show as in Fig. 9.1 that $m$ can also be written as

$$m=\left|\frac{d_i}{d_o}\right| \tag{9.4}$$

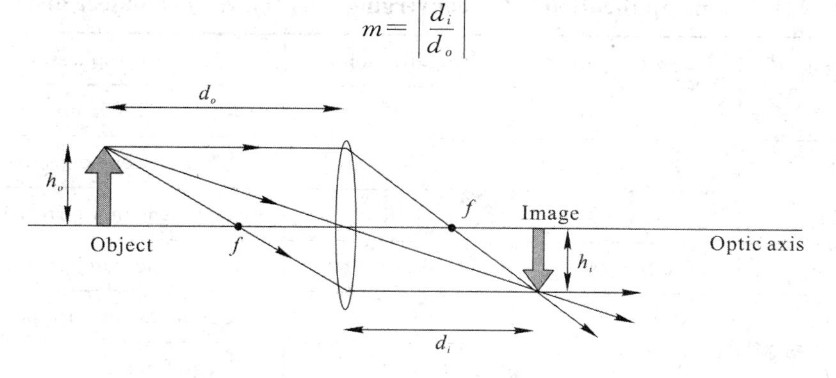


Fig. 9.1 The lateral magnification.

A converging lens is one which is thicker at the center than at the periphery, and converges incident parallel rays to a real focus on the opposite side of the lens from the object. A diverging lens is thinner at the center than at the periphery and diverges the light from a virtual focus on the same side of the lens as the object.

Fig. 9. 2 shows the formation of a real image by a converging lens. If the object distance is less than the focal length, then the image is virtual as shown in Fig. 9. 2(a) Some applications of a converging lens for various object distance $d_0$ ate given in Table 9. 1. The corresponding image distance $d_i$ follows Equation (9. 2) and the magnification $m$ follows from Equation(9. 3). Anegative value of $m$ means the image is inverted and $|m|>1$ implies it is enlarged.

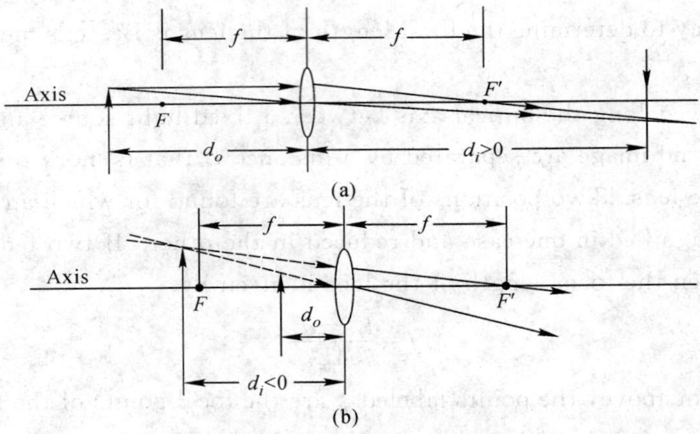


Fig. 9. 2 The formations of the images by a conversing lens for various object distance $d_o$.

**Table 9. 1 Some applications of a converging lens for various object distances.**

| Object distance $d_o$ | Image distance $d_i$ | Magnification $m$ | Application |
|---|---|---|---|
| $\infty$ | $f$ | 0 | Solar furnace |
| $d_o>2f$ | $2f<d_i<f$ | $d_o>m>-1$ | Telescope objective lens |
| $d_o=2f$ | $d_i=2f$ | $-1$ | Inverter lens in a terrestrial telescope |
| $2f>d_o>f$ | $d_i>2f$ | $<-1$ | Slide projector lens |
| $d_o=f$ | $\infty$ | $-\infty$ | Collimator to form parallel rays |
| $d_o<f$ | $0>d_i>-\infty$ | $>1$ | Magnifying glass, eyepiece lens |

A diverging lens is thinner at the center than at the periphery and diverges the light from a virtual focus on the same side of the lens as the object. A diverging lens makes rays diverge and by itself it can only form virtual images of real objects. Rays are drawn for two cases in Fig. 9.3.

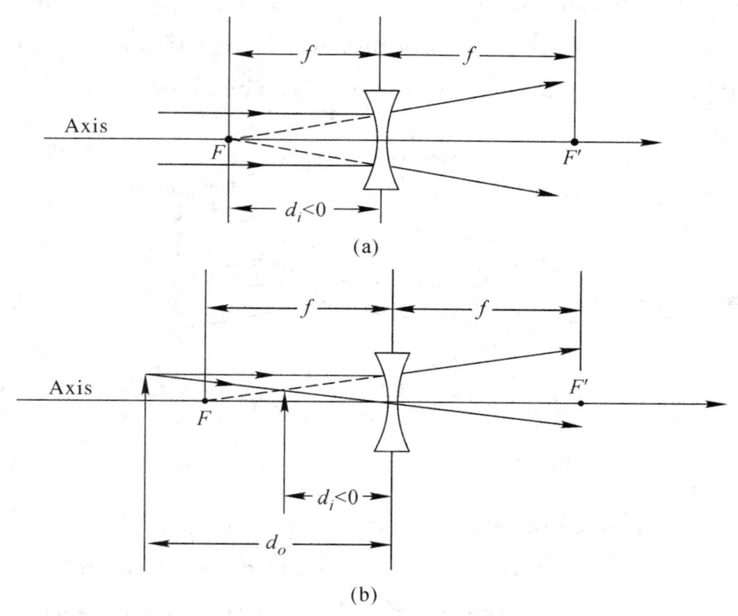


Fig. 9.3 The formations of the image by a diverging lens for various object distance $d_o$.

The diagram also shows you how a camera works: You have an object on one side of the lens and you have film or (for digital cameras) a CCD or CMOS imaging array on the other side. You focus the camera by adjusting the spacing between the lens and the film. Most cameras have multiple lenses to allow them to produce excellent images with short space between the lens and the film, but the principle of operation is as above.

## Procedure

1. Measurment of the focal length of converging lenses

There are three main methods to measure the focus length of converging lenses.

(1) Method 1: Use the lens to form an image of an object, measure the distances $d_i$ and $d_o$, and then use Equation (9.2) to compute $f$.

(2) Method 2: By using additional plane mirror.

The plane mirror is placed behind the lens and a position of coincidence of image and object in front of the lens is found. As object and image coincide the light from the object must have traveled to and from the mirror along the same path. This is only possible if the light reaches the mirror normal to its surface, i. e. , after reflection by the mirror it returns as a parallel beam of light. This parallel beam suffers refraction as it returns through the lens and is brought a focus in focal plane of the lens. (See Fig. 9. 4)

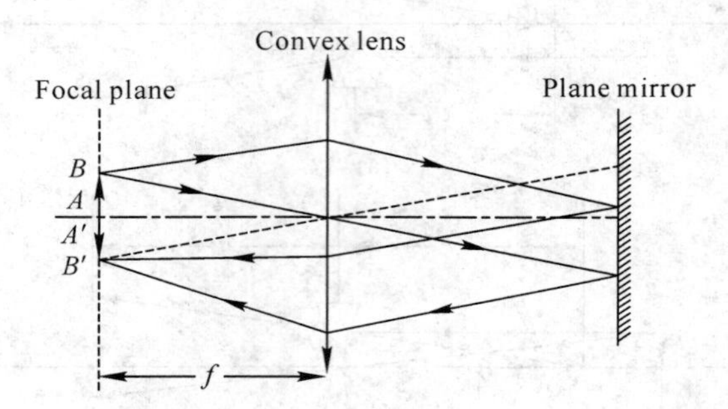


Fig. 9. 4 Focus length of coincidence.

(3) Method 3: Focal length by conjugate-foci method.

Mount the illuminated object near on end of the meter stick and place the screen at some convenient distance toward the other end, such that the distance between them is about 60 cm more than $4f$. Adjust the lens at the two positions for image formation, and determine whether both the enlarged and the diminished images are clear and distinct. If not, change the distance between the screen and the object until they are, and then record their positions. (See Fig. 9. 5)

Now make three independent trials for each position of lens and record the position of the lens, the object, and the screen, together with the image distances for each. Some time may be saved by alternating positions between trials.

From the means of the values for $d$ and $L$, compute the focal length of the lens by Equation (9. 3), making the computations a part of your report. Determine the values of $L$ and $d$, compute the focal length from Equation (9. 3) and include it in your report.

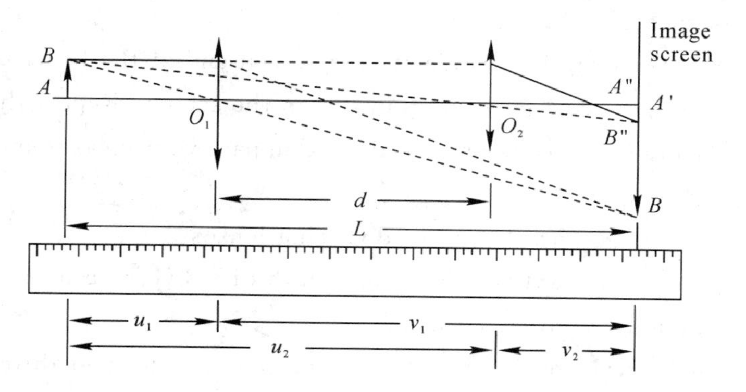


Fig. 9.5 Focal length by conjugate-foci method.

2. Images and magnification

In this part, you will use Method I to measure the focal length of large and small lenses.

(1) Place the light source at the end of the optics bench and attach it with the thumbscrew in the slot. Place the arrow aperture on the front of the light source. It will save a little trouble in your calculations if you position the source so that the object (the frosted arrow) is exactly beside an integer mark (e. g. 5. 0 cm) on the scale of the bench. Gently tighten the thumbscrew to secure the source, and record the position of the object. Connect the light source to the power supply and momentarily depress the "start" switch to turn on the light. Place the frosted screen at the far end of the bench and note its position, as indicated by the ring inscribed on the housing. Again, it will save some trouble if you locate it a convenient integer mark, like 90. 0 cm or 92. 0 cm.

Now put the lens in the foam holder on the bench close to the light source and move it slowly away from the source until you see a clear image on the screen. The image is most easily seen looking through the screen towards the light source, but it can also be seen from the other side. Adjust the position of the lens to give the sharpest image and record the position of the lens (you will need to measure how far inside the foam housing your lens is positioned, etc. ). Calculate $d_o$, $d_i$, and from Equation (9. 1), the focal length, $f$, of the lens.

(2) If the image is not centered on the screen, adjust the position of the object plate on the front of the light source until the image is centered. Now measure $h_o$ and $h_i$, the heights of the object and image. Compute the lateral magnification $m = |h_i/h_o|$ and

compare with the expected value $|d_i/d_o|$.

(3) Next, move the foam lens holder to the screen end of the bench and then slide it away from the screen until you get a sharp image on the screen. Repeat the measurements above for $d_o$, $d_i$, $h_i$ and $h_o$. Recompute $f$ and $m$. Compare your measurements in Parts (2) and (3).

3. Measurement of the focal length of diverging lenses

Here two measurement methods for focus length of a diverging lens are introduced.

(1) By using auxiliary converging lenses.

As in Fig. 9. 6, a real image $A'B'$ of the object $AB$ is formed through an auxiliary converging lens. Then adding the diverging lens under measurement, a real image $A''B''$ is formed. However, for the diverging lens under measurement $A'B'$ and $A''B''$ are a virtual object and a real image, respectively. If we know the distances from the diverging lens $L_2$ to $A'B'$; plane and $L_2$ to $A''B''$ plane, the focal length of the diverging lens $f_2$ can be calculated by Equation (9. 2). We have

$$\left.\begin{aligned} u_2 &= v_1 - d \\ \frac{1}{f_2} &= \frac{1}{u_2} + \frac{1}{v_2} \end{aligned}\right\} \tag{9.5}$$

where $v_1$ is the distance of the image from the auxiliary converging lens, $u_2$ is the distance from the object from the diverging lens, and $d$ is the distance between the converging lens and the diverging lens.

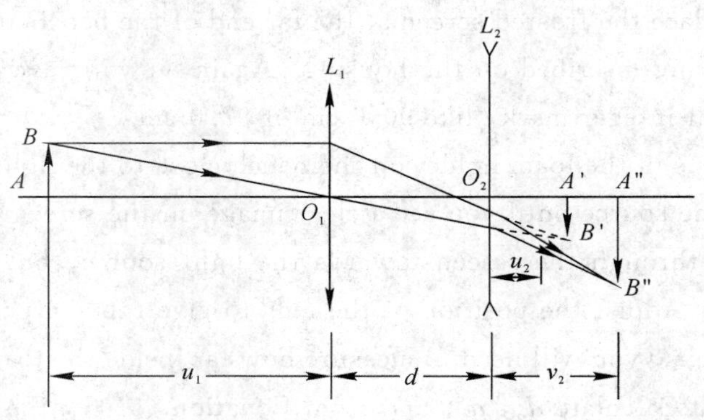


Fig. 9. 6 Focal length measurement by using auxiliary converging lens.

(2) By using additional plane mirror.

The theory of the measurement of the focal length for a diverging lens by using additional plane mirror is the same as for a converging lens.

## Observations and Results

1. Focal length by direct measurement. Mount a converging lens on the optical bench, together with a white screen. By using a distant brightly illuminated object, as viewed through a window, make a trial to locate the position of the real image thus formed, record the object distance and the image distance, determine the focal length of the converging lens (Table 9.1).

**Table 9.1 The data for measuring the focal length of a converging lens directly.**

<table>
<tr><th>$n$</th><th>Object position $x_0$/cm</th><th>Lens position $x_L$/cm</th><th>Image position $x_i$/cm</th><th>$f$/cm</th></tr>
<tr><td>1</td><td></td><td></td><td></td><td rowspan="4">$f=\frac{d_o d_i}{d_o+d_i}=$</td></tr>
<tr><td>2</td><td></td><td></td><td></td></tr>
<tr><td>3</td><td></td><td></td><td></td></tr>
<tr><td colspan="4">$d_o=|x_o-x_L|$; $d_i=|x_i-x_L|$</td></tr>
</table>

2. Focal length by coincidence. Select a converging lens. Mount a plane mirror from the lens. Mount the illuminated object on the opposite side of the lens and adjust its position until a distinct image of the cross wire shows on the object screen just beside the object. The light is now returning nearly along its original path. This requires that the rays be parallel on the opposite side of the lens from the object, and that they strike the mirror perpendicular to its surface. The point at which the parallel rays form an image is the principal focus. The object and image are nearly coincident and their distance from the lens is the focal length. Record the image distance and determine the mean value of the focal length (Table 9.2).

**Table 9.2 The data for measuring the focal length of a converging lens by coincidence.**

| $n$ | Object position $x_o$/cm | Lens position $x_L$/cm | Focal length of the converging lens $f$/cm<br>$f=\|x_L-x_o\|$ |
|---|---|---|---|
| 1 | | | |
| 2 | | | |
| 3 | | | |
| Mean | | | |

3. Focal length by conjugate-foci method. Select a converging lens. Mount the illuminated object near one end of the optical bench, and place the screen at some convenient distance toward the other end so that the distance between them is more than $4f$. Adjust the lens at the two positions for image formation, and determine whether both the enlarged and the diminished images are clear and distinct. If not, change the distance between the screen and the object until they are distinct, and then record their positions (Table 9.3).

**Table 9.3 The data for measuring the focal length of a converging lens by conjugate-foci method.**

| $n$ | Object position $x_o$/cm | Image position $x_i$/cm | Lens position $x_{L1}$/cm as enlarged image | Lens position $x_{L_2}$/cm as diminished image | Focal length of the converging lens $f$/cm |
|---|---|---|---|---|---|
| 1 | | | | | $L=\|x_o-x_i\|$<br>$d=\|x_{L1}-x_{L2}\|$<br>$f=\frac{L^2-d^2}{4L}=$ |
| 2 | | | | | |
| 3 | | | | | |
| mean | | | | | |

## Questions

1. Consider a lens with a focal length of $f=20.0$ cm which is used to image an object of height $h_o=4.0$ cm, a distance $d_o=40.0$ cm away. On graph paper, draw a diagram showing the size ($h_i$) and position ($d_i$) of the image formed by this lens. Check if the value of $d_i$ obtained from your graph agrees with a calculation of $d_i$ by Equation (9.1).

2. What is a collimated beam? Draw a sketch showing how you could produce one. Could you make a collimated beam using only a point source and a diverging lens?

3. The Earth's moon as seen from the surface of the Earth has an angular size of roughly 0.5 degrees. We say that it 'subtends' an angle of 0.5 degrees. What is this angle in radians? The radian angular measurement is convenient because (by definition) the angle in radians is the ratio of the length of arc subtended to the radius of the circle.

# Lab 10 Light Polarization—Malus's Law

## Purpose

To observe the transmission of linearly polarized light through a polarizer.

## Apparatus

Laser diode; polarizer; slide rail; photodiode; electricity meter.

## Theory

Light is an electromagnetic wave, and the polarization direction of an electromagnetic wave is defined as the vibration direction of the electric field for the wave. In a vacuum or uniform medium, the direction of polarization is perpendicular to the propagation direction of an electromagnetic wave. As shown in Fig. 10.1(a), an electromagnetic wave propagates along the $x$-axis, and the polarization direction follows the $y$-axis. The direction of polarization is commonly represented as the angle between the vibration direction of the electric field and a reference direction in a plane perpendicular to the propagation direction of the electromagnetic wave. As shown in Fig. 10.1(b), if the $y$-axis is the reference direction, the wave has a polarization angle of 0°.

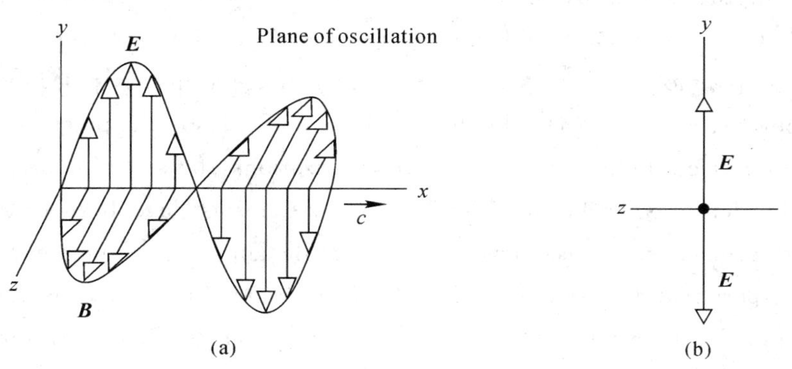


Fig. 10.1(a) The correlation between the electric field **E** and magnetic **B** of an electromagnetic wave.
(b) The vibration direction of **E** in a perpendicular plane.

However, for any light wave (or electromagnetic wave) commonly found in nature, the polarization direction of the electric field exists in every angle. As shown in Fig. 10.2(a), one light beam represents a superimposition of electric fields in various directions. The result is an "unpolarized" electromagnetic wave, such as sunlight or light emitted from an incandescent bulb. If the electric field of a given light ray vibrates in a certain direction at all times, this ray or beam is considered "linearly polarized light." The term "linearly polarized" indicates that when observing the propagation direction of linearly polarized light, the electric field vibrates along a line over time (Fig. 10.2(b)).

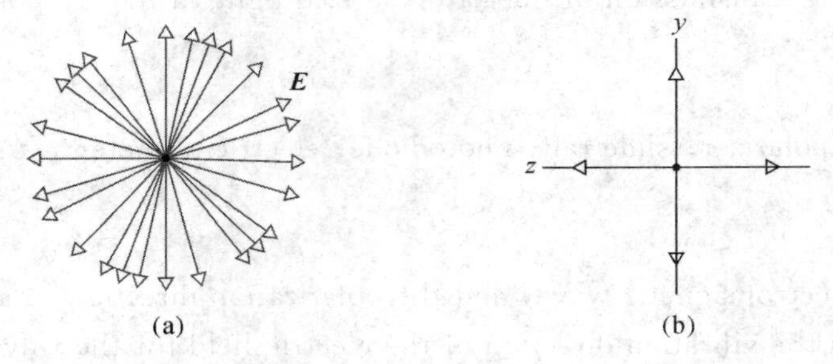


Fig. 10.2(a) Vibration direction of the electric field $\boldsymbol{E}$ of an unpolarized light ray.
(b) Representation of any electric field $\boldsymbol{E}$ using two mutually perpendicular components.

Employing vectors for analysis, we find that the electric field of an electromagnetic wave consists of two components that are mutually perpendicular and possess the same frequency and wavelength. In the case of a linearly polarized wave, these two components of the electric field must have a phase difference of 0° or 90°. An electromagnetic wave is circularly or elliptically polarized if the phase difference between these two components is not 0° or 90°. Consequently, the direction of the electric field varies over time.

Polarizers are among the tools widely used to produce a linearly polarized light ray or beam. As shown in Fig. 10.3(a), this tool passes light of a specific polarization direction, and absorbs or reflects light of perpendicular polarizations, by selecting a specific electric field of the incident ray or beam. Varying polarizer forms have been created to fulfill various design principles and accommodate different demands. The common polarizer types used in this experiment (as shown in Fig. 10.3(b)) are represented as the "polarizing sheet" shown in Fig. 10.3(a), where vertical lines represent the vibration directions of an electric field that are allowed to pass the polarizer.

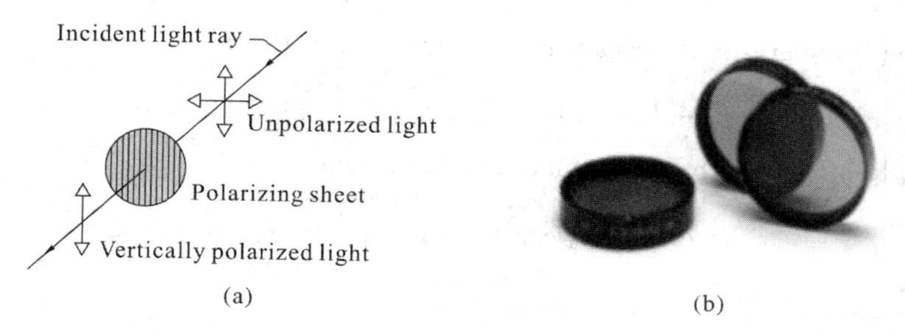


Fig. 10.3(a) The vibration direction of unpolarized light transmitted through a polarizing sheet. (b) Common polarizers.

Fig. 10.4 shows an unpolarized light ray passing through two sequential polarizes. Each polarizer has a transmission axis that designates a direction, and the polarization directions of electric fields parallel to the designated direction can be transmitted through the polarizer. The transmission axis of the second polarizer is set at an angle of $\theta$ to the transmission axis of the first polarizer.

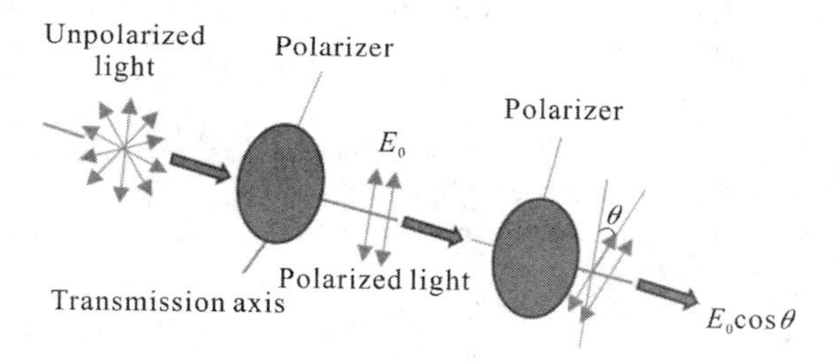


Fig. 10.4 Schematic diagram of Malus's law.

The unpolarized ray is linearly polarized by the first polarizer. Additionally, we assume that the electric field amplitude of the polarized ray is $E_0$. The amplitude of the electric field after the ray is transmitted through the second polarizer will be $E_0\cos\theta$ because angle $\theta$ exists between the transmission axis of the second polarizer and the polarization direction of the incident ray.

The brightness of a ray or beam is related to its power density, which is the power per unit time per unit area transmitted by the ray (or electromagnetic wave). The power transmitted by an electromagnetic wave is represented using the Poynting vector as $\boldsymbol{S}=\frac{1}{\mu_0}\boldsymbol{E}\times\boldsymbol{B}$. During experiments, light is measured as the time-averaged Poynting vector $\boldsymbol{S}_{avg}$, which is known as the intensity of light ($I$).

Thus, the intensity of the ray transmitted through the first polarizer is $I_0 = \frac{c\varepsilon_0{}^2}{2} E_0^2$. After the ray passes through the second polarizer, which has a transmission axis that creates an angle of $\theta$ with the first polarizer, the intensity is

$$I(\theta) = \frac{c\varepsilon_0^2}{2} E_0^2 \cos^2\theta \quad \text{or} \quad I(\theta) = I_0 \cos^2\theta \tag{10.1}$$

This is Malus's law.

Three polarizers

Unpolarized light passes through 3 polarizers (See Fig. 10.5):

The first and last polarizers are oriented at 90° with respect to each other. The second polarizer has its polarization axis rotated an angle $\phi$ from the first polarizer. Therefore, the third polarizer is rotated an angle $\frac{\pi}{2} - \phi$ from the second polarizer. The intensity after passing through the first polarizer is $I_1$ and the intensity after passing through the second polarizer, $I_2$, is given by $I_2 = I_1 \cos^2\phi$. The intensity after the third polarizer, $I_3$, is given by:

$$I_3 = I_2 \cos^2\left(\frac{\pi}{2} - \phi\right) = I_1 \cos^2\phi \cos^2\left(\frac{\pi}{2} - \phi\right) \tag{10.2}$$

Rearranging Equation(10.2), we obtain

$$I_3 = \frac{I_1}{4} \sin^2(2\phi) \tag{10.3}$$

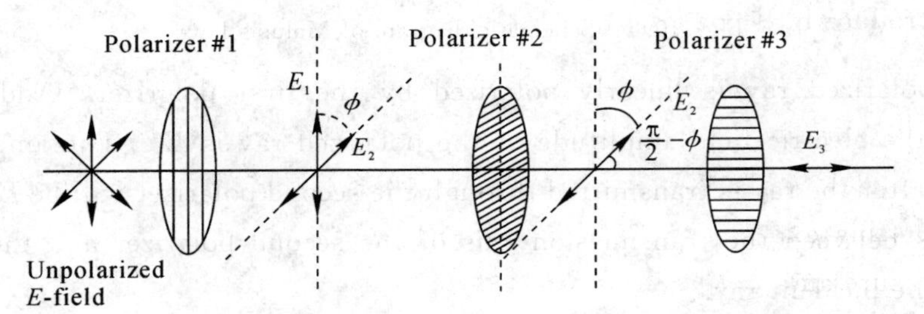


Fig. 10.5 The electric field transmitted through three polarizers.

Because the data acquisition begins when the transmitted intensity through Polarizer 3 is a maximum, the angle $\theta$ measured in the experiment is zero when the second polarizer is 45° from the angle $\phi$. Thus the angle $\phi$ is related to the measured angle $\theta$ by:

$$\phi = 45° + \theta \tag{10.4}$$

## Procedure

### A. Instrument Setup

Fig. 10. 6 shows the experimental setup. The relevant details are provided below.

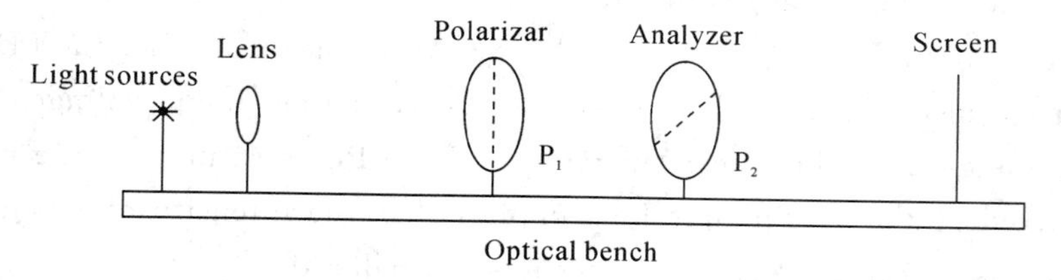


Fig. 10. 6 Producing and analyzing of the plane polarized light.

1. Component positions and heights

The light and photo diode (PD) are separately placed on the slide platforms on the outermost slide bases of the slide rail.

Use the height of the light beam as the standard for other components. Align the components and ensure that the beam hits the center of the PD detector through the centers of the three polarizers. The direction of the beam should be parallel to the slide rail.

2. PD setup

Connect the two voltage output cables to the multimeter. The black cable is the ground cable, and the red cable is used for voltage signals. Set the electricity meter to DC.

The measured light intensity is output by the PD in analog voltage signals. Higher intensities produce higher voltages.

### B. Measuring the Polarizations of Components

In this experiment, components such as the laser and polarizers have specific polarization angles. Polarizers # 1 and # 2, which have an adjustable angle, should be employed for this experiment. Use the angle graduations or increments on the support frame to measure and record the polarization direction of the light beam and the directions of the transmission axes of the other polarizers. The measurement procedures for the above instruments are as follows:

1. Ensure that the beam hits the detector in the PD.

2. The relative positions of the components can be located using the beam propagation direction (Fig. 10. 4).

3. Test and adjust the intensity of the beam illuminating the PD. Changes in intensity

must be measured accurately.

(1) If the beam is too intense, the PD becomes saturated and cannot correctly measure the voltage for higher light intensities.

(2) If the beam is too weak, the PD and electricity meter measurements will be too imprecise to reflect minor changes in intensity.

(3) Rotate Polarizer #1 (approximately 180°) and assess whether the PD is saturated by observing the meter reading. In addition, ob serve the maximum voltage.

When the angle is within a certain range and the PD is saturated, the meter reading will remain fixed at the maximum. This implies that the intensity of the incident beam projected into the detector is excessive and must be reduced.

(4) Tape a piece of paper or a thin black film to the front of the detector to reduce the intensity of the beam illuminating the detector.

4. Record the background signal measured by the PD.

Block the laser beam to enable the PD to measure only the light within the room. Subtract the background signal from the measured light intensity for correction.

5. Alter the angle of Polarizer #2 and record the electricity meter readings measured at various angles.

Let the angle be represented as a horizontal axis and output voltage be represented as a vertical axis, and plot a data graph. The range of angle alteration should be approximately 360°.

6. According to the data graph, calculate the polarizations of the light and the angles of the transmission axes for Polarizers #1 and #2.

## Questions

1. Compare your plot of the data to your plot of the fitted curve. How well do your results agree with Malus's Law? Explain why it agrees or does not agree.

2. How well does your measurement of Brewster's angle agree with the value you would expect if the glass has $n=1.50$? State clearly what you measured and what you expected and then explain why your result agrees or does not agree.

3. Give two possible reasons why the reflected light intensity does not quite reduce to zero at Brewster's angle in the apparatus you used today. Explain your answers briefly.

# Lab 11 The Michelson Interferometer

## Purpose

1. To measure the wavelength of light from a Ne-He laser use a Michelson interferometer;
2. To measure the index of refraction of glass.

## Apparatus

Michelson interferometer; laser light source; thin glass.

## Theory

Interferometers are used to precisely measure the wavelength of optical beams through the creation of interference patterns. The Michelson interferometer is a historically important device which provides simple interferometer configuration, useful for introducing basic principles.

1. Intervene Theory

Light is a transverse wave. When two waves of same wavelength and amplitude travel through the same medium, their amplitudes combine. A wave of greater or lesser amplitude than the original will be the result. The addition of amplitudes due to superposition of two waves is called interference. If the crest of one wave meets with the trough of the other, the resultant intensity will be zero and the waves are said to interfere destructively. Alternatively, if the crest of one wave meets with the crest of the other, the resultant will be maximum intensity and the waves are said to interfere constructively.

Suppose two coherent (*i. e.*, their initial phase relationship remains constant) waves start from the same point and travel different paths before coming back together and interfering with each other. Suppose also that the re-combined waves illuminate a screen where the position on the screen depends on the difference in the path lengths traveled by the two waves. Then the resulting alternating bright and dark bands on the screen are called interference fringes.

In constructive interference, a bright fringe (band) is obtained on the screen. For constructive interference to occur, the path difference between two beams must be an integral multiple $m\lambda$ of the wavelength $\lambda$, where $m$ is the order, with $m=0$, 1, 2… If the path difference between two waves is $(m+1/2)\lambda$, the interference between them is destructive, and a dark fringe appears on the screen.

2. Michelson Interferometer

The Michelson interferometer is the best example of what is called an amplitude-splitting interferometer. It was invented in 1893 by Albert Michelson, to measure a standard meter in units of the wavelength of the red line of the cadmium spectrum. With an optical interferometer, one can measure distances directly in terms of wavelength of light used, by counting the interference fringes that move when one or the other of two mirrors are moved. In the Michelson interferometer, coherent beams are obtained by splitting a beam of light that originates from a single source with a partially reflecting mirror called a beam splitter. The resulting reflected and transmitted waves are then re-directed by ordinary mirrors to a screen where they superimpose to create fringes. This is known as interference by division of amplitude. This interferometer, used in 1817 in the famous Michelson-Morley experiment, demonstrated the non-existence of electromagnetic-wave-carrying ether, thus paving the way for the Special Theory of Relativity.

3. How the Interferometer Works

A simplified diagram of a Michelson interferometer is shown in Fig. 11.1. Light from a monochromatic source S is divided by a beam splitter (BS), which is oriented at an angle 45° to the beam, producing two beams of equal intensity. The transmitted beam (T) travels to mirror $M_1$ and it is reflected back to BS. 50% of the returning beam is then reflected by the beam splitter and strikes the screen, E. The reflected beam (R) travels to mirror $M_2$, where it is reflected. 50% of this beam passes straight through beam splitter

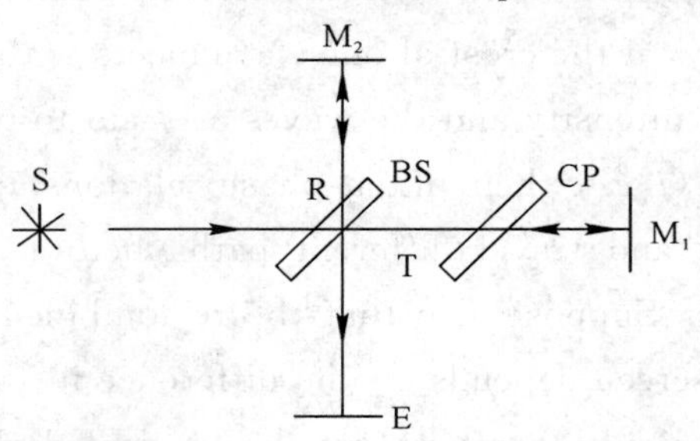


Fig. 11.1 A simplified diagram of a Michelson interferometer.

and reaches the screen.

Since the reflecting surface of the beam splitter BS is the surface on the lower right, the light ray starting from the source S and undergoing reflection at the mirror $M_2$ passes through the beam splitter three times, while the ray reflected at $M_1$ travels through BS only once. The optical path length through the glass plate depends on its index of refraction, which causes an optical path difference between the two beams. To compensate for this, a glass plate CP of the same thickness and index of refraction as that of BS is introduced between $M_1$ and BS. The recombined beams interfere and produce fringes on the screen E. The relative phase of the two beams determines whether the interference will be constructive or destructive. By adjusting the inclination of $M_1$ and $M_2$, one can produce circular fringes, straight-line fringes, or curved fringes. This lab uses circular fringes, shown in Fig. 11.2.

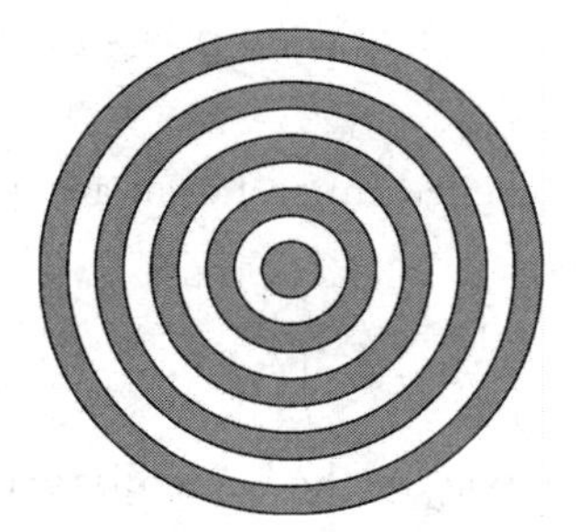

Fig. 11.2 Circular fringes.

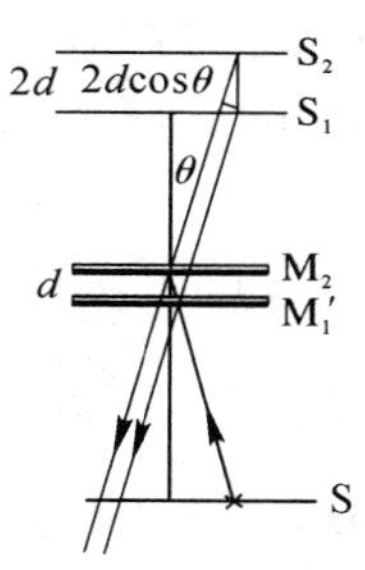


Fig. 11.3 The measuring schematic.

From the screen, an observer sees $M_2$ directly and the virtual image $M_1'$ of the mirror $M_1$, formed by reflection in the beam splitter, as shown in Fig. 11.3. This means that one of the interfering beams comes from $M_2$ and the other beam appears to come from the virtual image $M_1'$. If the two arms of the interferometer are equal in length, $M_1'$ coincides with $M_2$. If they do not coincide, let the distance between them be $d$, and consider a light ray from a point S. It will be reflected by both $M_1'$ and $M_2$, and the observer will see two virtual images, $S_1$ due to reflection at $M_1'$, and $S_2$ due to reflection at $M_2$. These virtual images will be separated by a distance $2d$. If $\theta$ is the angle with which the observer looks into the system, the path difference between the two beams is $2d\cos\theta$. When the light that comes from $M_1$ undergoes reflection at BS, a phase change of $\pi$ occurs, which corresponds to a path difference of $\lambda/2$. Therefore, the total path difference between the two beams is $\Delta = 2d\cos\theta + \frac{\lambda}{2}$. The condition for constructive interference is then,

$$\Delta=2d\cos\theta+\frac{\lambda}{2}=m\lambda, \qquad m=0, 1, 2 \cdots \tag{11.1}$$

For a given mirror separation $d$, a given wavelength $\lambda$, and order $m$, the angle of inclination $\theta$ is a constant, and the fringes are circular. They are called *fringes of equal inclination*, *or Haidinger fringes*. If $M_1'$ coincides with $M_2$, $d=0$, and the path difference between the interfering beams will be $\lambda/2$. This corresponds to destructive interference, so the center of the field will be dark.

If one of the mirrors is moved a distance $\lambda/4$, the path difference changes by $\lambda/2$ and a maximum is obtained. If the mirror is moved another $\lambda/4$, a minimum is obtained; moving it by another $\lambda/4$, again a maximum is obtained; and so on. Because $d$ is multiplied by $\cos\theta$, as $d$ increases, new rings appear in the center faster than the rings already present at the periphery disappear, and the field becomes more crowded with thinner rings toward the outside. If $d$ decreases, the rings contract, become wider and more sparsely distributed, and disappear at the center.

For destructive interference, the total path difference must be an integer number of wavelengths plus a half wavelength,

$$\Delta_{\text{destr}}=2d\cos\theta+\frac{\lambda}{2}=\left(m+\frac{1}{2}\right)\lambda, \qquad m=0, 1, 2, \cdots \tag{11.2}$$

If the images $S_1$ and $S_2$ from the two mirrors are exactly the same distance away, $d=0$ and there is no dependence on $\theta$. This means that only one fringe is visible, the zero order destructive interference fringe, where $m=0$ and the observer sees a single, large, central dark spot with no surrounding rings.

## Procedures

1. Measurement of wavelength

(1) Using the Michelson interferometer, the wavelength of light from a monochromatic source can be determined. If $M_2$ is moved forward or backward, circular fringes appear or disappear at the centre. The mirror is moved a known distance $d$ and the number $N$ of fringes appearing or disappearing at the centre is counted. For one fringe to appear or disappear, the mirror must be moved a distance of $\lambda/2$. Knowing this, we can write $d=N\lambda/2$, So that the wavelength is

$$\lambda=\frac{2d}{N} \tag{11.3}$$

(2) Adjustments

① The laser beam must strike at the center of the movable mirror and should be reflected directly back into the laser aperture.

② Adjust the position of the beam splitter so that the beam is reflected to the fixed mirror.

③ Adjust the angle of beam splitter to be 45 degrees. There will be two sets of bright spots on the screen, one set from the fixed mirror and the other from the movable mirror.

④ Adjust the angle of the beam splitter to make the two sets of spots as close together as possible.

⑤ With the screws on the back of the adjustable mirror, adjust the mirror's tilt until the two sets of spots on the screen coincide.

⑥ Expand the laser beam slowly by rotating the collimating lens in front of the laser.

⑦ Align the laser with the interferometer and make certain that the fringes are moving when the micrometer screw is turned.

⑧ Mark a point on the screen and note the micrometer reading.

⑨ As the screw is moved, the fringes begin to displace. Count the number of fringes $N$ that move past the mark (either inward or outward). To avoid the effects of backlash in the micrometer screw, turn the micrometer handle one full turn before starting the count.

⑩ Note the micrometer readings at the beginning and end of the count. Calculate the distance $d'$ the mirror is moved, according to the beginning and ending micrometer readings. Repeat the procedure several times. Average the readings.

⑪ With a known wavelength laser, use $d = N\lambda/2$ to calculate the actual distance moved. The calibration constant of the interferometer is then $k = d/d'$. All subsequent distance measurements with the micrometer should be multiplied by the calibration constant $k$. Ideally, $k$ would be exactly 1, but factors such as wear and thermal expansion can cause it to vary.

⑫ Once the calibration constant is known, if the laser source has an unknown wavelength, it can be calculated with the same equation.

2. Observation of white light fringes

Because of variable wavelength composition and short coherent length, just a few white light fringes could occur as the movable mirror $M_2$ is superposed to the fixed image $M_1'$.

The adjustments for laser light should be well made at first. The fringes should appear as a series of concentric circles by careful adjustment of the moveable mirror $M_1$ parallel to the image $M_1'$.

The second step is to access and superpose the movable $M_2$ to the fixed $M_1'$, It would sharpen the circular profiles wider and decrease their amount to the smallest, download the screen and change the laser to a white light source. Turn the minute wheel carefully until the circular fringes appear again in front of the movable mirror $M_2$ behind the splitter $G_1$. The white light fringes would be composed to a white or black at the center with a few colors around.

3. Index of refraction of glass plate

A thin glass plate should be mounted in a holder on the interferometer between the beam splitter and the fixed arrow. The holder must be capable of a slow rotation through a measured angle.

The instrument is aligned to produce circular fringes of a monochromatic light. The glass plate is then rotated through an angle sufficient to produce a shift of about one hundred fringes. This shift in fringes is caused by the increase in optical path due to the index of refraction of the glass. The fringe shift (number of fringes) multiplied by the wavelength is equal to twice the increase in optical path length, $n\Delta x$. From the measurement of the angle through which the glass plate is rotated, the fringe shift and the thickness of the glass plate, its index of refraction can be calculated.

A raw way to measure the refraction index of glass is based on a white light source. The instrument is at first aligned with the movable mirror $M_2$ on the primary site $D_0$ and produces circular white light fringes. And then a glass plate is placed in front of the movable mirror $M_2$. Adjust the minute wheel until the circular fringes appear again with the mirror on the site $D_1$. From the measurement of the replacement of $M_2$ and the thickness $L$ of the glass plate, the refraction index can be estimated by

$$n=\frac{d}{L}+1, \quad d=D_1-D_0 \tag{11.4}$$

4. Measurement of sodium doublet

Sodium doublet light has two series interferential fringes for wavelengths $\lambda_1=589.0$ nm and $\lambda_1=589.6$ nm. The bright fringes of $\lambda_1$ would become dull because of superposing the blacks of $\lambda_1$, as the optical path differences of the two coherent beams are equal to

$$(K+nK')\lambda_1=\left(K+nK'+\frac{2n-1}{2}\right)\lambda_2 \tag{11.5}$$

From the measurement of the replacement $\Delta D$ of $M_2$ for two neighboring dullest bright fringes, the difference of the two wavelengths can be estimated as

$$\Delta\lambda = \frac{\lambda_1 \lambda_2}{2\Delta D} \tag{11.6}$$

For a monochromatic wavelength $\lambda$ and wavelength width $\Delta\lambda$, the interferential fringes would disappear for the dulling effect when the optical path difference of two coherent beams is lager than a so-called length

$$L_m = \frac{\lambda^2}{\Delta\lambda} \tag{11.7}$$

Time spent through the path $L_m$ with speed of light $c$ is called coherent time, that is

$$t_m = \frac{L_m}{c} \tag{11.8}$$

For laboratory optical sources, white beam's coherent lengths are very short, laser's being long of meters or even kilometers, and sodium doublet's about 2 cm. The coherent length of sodium doublet can be determined by the replacement of movable mirror $M_2$ from the site of zero optical path difference to that of fringes disappearing.

## Questions

1. Compute the coherence time and coherence length for white light, which may be modeled as containing "equal amounts" of light with all frequencies from the blue (400 nm) to the red (700 nm).

2. Redraw the optical configuration of the Michelson to show the separation of the effective point sources due to the path-length difference in the two arms of the interferometer.

3. Explain the direction of motion of the circular fringes when the path length is changed, i. e. , what directions do the circular fringes move if the OPL is increased? What if OPL is decreased?

4. When the Michelson is used with collimated light, explain how a single dark fringe can be obtained. Where did the light intensity go?

5. Explain what will happen when a piece of glass (or other material) is placed in one arm.

# Lab 12 Diffraction and Interference of Plane Light Waves

## Purpose

1. To investigate the diffraction patterns produced by monochromatic laser light for various slit configurations;

2. To determine the wavelength of the laser light from measurements of the intensity distribution of light for the single slit;

3. To determine the thickness of a human hair strand from the diffraction pattern.

## Apparatus

He-Ne laser; single slit move with holder; aperture disk; light sensor interface device; optical bench.

## Theory

When coherent light of monochromatic wavelength is incident upon a slit, the light diverges as it passes through the slit in a process known as diffraction. A laser produces coherent light, which means all the light striking the slit is in phase. If the light then falls on a screen placed at a large distance from the slit, it produces a pattern of alternating bright and dark images of the slit. This pattern is referred to as a Fraunhofer diffraction pattern, which is the simplest case of diffraction. It occurs when rays emerging from the slit can be considered to be parallel.

1. Single-slit interference

The diffraction process is explained by the fact that light is a form of electromagnetic wave and the different portions of the slit behave as if they were separate sources of light waves—Huygens' principle. At each point on the screen, the light from different portions of the slit will have a different phase due to the different path lengths of light from each portion of the slit to the point on the screen. Light from the different portions of the slit will interfere with each other and the resultant intensity will vary at different places on the screen. This is illustrated in Fig. 12.1 where five rays are shown from different portions of the slit. The choice of five is arbitrary and the slit could be divided into any number of portions.

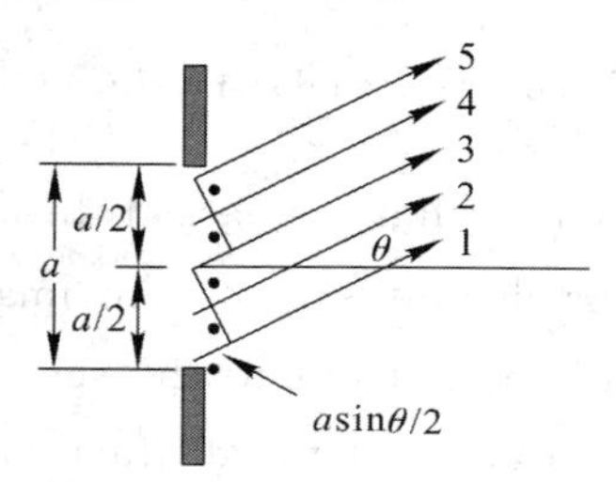


Fig. 12.1 Path of five rays from different portions of the slit.

Consider Wave 1 from near the bottom of the slit and Wave 3 from the center of the slit. They differ in path length by the amount $(a\sin\theta)/2$ as shown, where $a$ is the width of the slit and $\theta$ is the angle each of the rays makes with the horizontal. It is also true that the path difference between rays 2 and 4 is $(a\sin\theta)/2$. If this path difference is exactly equal to half a wavelength, i. e. , $\lambda/2$, the two waves will be 180° out of phase and will interfere destructively and cancel each other. Therefore, waves from the upper half of the slit will be 180° out of phase with waves from the lower half of the slit when

$$\frac{a}{2}\sin\theta=\frac{\lambda}{2} \tag{12.1}$$

By dividing the slits into $2m$ portions, it can be shown that the condition for destructive interference will be satisfied at angle $\theta$ on the screen above and below the center of the pattern given by:

$$\sin\theta=m\frac{\lambda}{a},\quad m=\pm1,\ \pm2,\ \pm3,\ \cdots \tag{12.2}$$

There is no simple expression for the location of the maxima on the screen other than that for the principal maximum at the center of the pattern. The other maxima are much less intense than the principal maximum and are located approximately halfway between the minima. The diffraction pattern that appears on a screen will have an intensity variation above and below the slit as shown in Fig. 12.2.

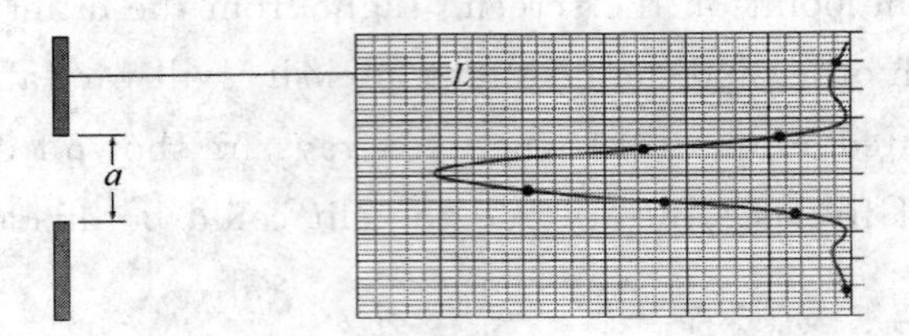


Fig. 12.2 Diffraction pattern on a screen distance $L$ from a single slit with width $a$.

To study diffraction of light, laser light is passed through a narrow single-slit and the diffraction pattern is formed on a distant screen. An imaginary reference line is drawn perpendicularly from the center of the slit out to the screen (see Fig. 12.3), which is a distance $L$ away. The intensity variation of the diffraction pattern can then be measured accurately as a function of the distance $x$ from the reference line. In the theoretical description of the diffraction pattern, however, it is more convenient to quantify the light intensity as a function of the sine of the angle $\theta$ defined accordingly by

$$\sin\theta = \frac{y}{\sqrt{y^2 + L^2}} \tag{12.3}$$

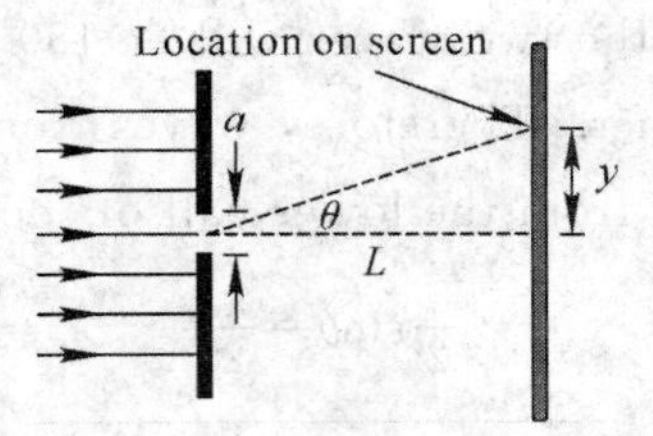


Fig. 12.3 A schematic diagram for the light diffraction setup.

The theory of diffraction predicts that the spatial pattern of light intensity on the viewing screen by a light wave passing through a single rectangular-shaped slit is given by

$$I(\theta) = I_0 \left[ \frac{\sin\left(\frac{\pi a \sin\theta}{\lambda}\right)}{\frac{\pi a \sin\theta}{\lambda}} \right]^2 \tag{12.4}$$

The observed intensity distribution, given by Eq. (12.4) is displayed in Fig. 12.4 for different values of "$a$" in terms of the wavelength $\lambda$.

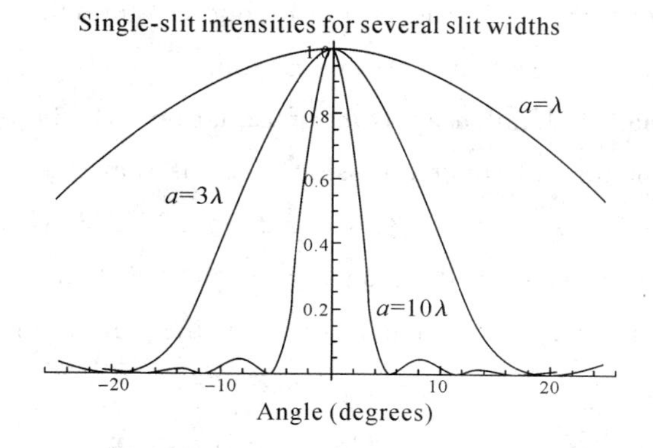


Fig. 12.4 Relative intensity vs. angle patterns for single-slit diffraction.

The intensity in the higher order maxima is much less than the central peak. Another feature is that with increasing slit width, the central peak becomes narrower and the secondary maxima more pronounced.

To understand the generic features on the intensity variation without having to make a particular choice for the slit width $a$, it is convenient to rewrite this expression in terms of the quantity $\alpha$:

$$\alpha \equiv \frac{a\sin\theta}{\lambda} \tag{12.5}$$

Then, Equation (12.4) becomes

$$I(\theta) = I_0 \left[\frac{\sin(\pi\alpha)}{\pi\alpha}\right]^2 \tag{12.6}$$

In the above expression, the angle of the sine function is in radians.

The plot of Equation (12.6) is shown in Fig. 12.5. It can be seen that the diffraction pattern formed when light wave passes through a rectangular-shaped slit consists of a set of bright spots (the principal maximum and many secondary maxima) interspersed with regions of darkness (the minima). A summary of the key features of this rectangular-slit diffraction pattern is as follows:

- **Minima.** The minima (locations of zero light intensity) occur at the angle $\theta$ given by $\alpha = a\sin\theta/\lambda = \pm 1, \pm 2, \pm 3, \pm 4\cdots$ and are called first, second, third, fourth, ⋯ minima, respectively. It is to be noted that the condition for minima is the same as that obtained in Eq. (12.2) from a simpler geometrical argument.

- **Principal Maxima.** The central peak, bracketed by the two "first minima" (which are located at $\alpha = a\sin\theta/\lambda = \pm 1$), is the region of highest light intensity and most of the diffracted wave's energy is concentrated in this region.

• **Secondary Maxima.** A detailed analysis of Equation (12.6) (which involves taking derivatives to find the maxima of this expression) reveals that the secondary maxima occur at angles $\theta$ given by $\alpha = a\sin\theta/\lambda = \pm 1.43030$, $\pm 2.45902$, $\pm 3.47089$, $\pm 4.47741$ and are called the first, second, third and fourth secondary maxima, and they are only 4.7%, 1.6%, 0.8%, 0.5% respectively of $I_0$ (the intensity of the principal maximum).

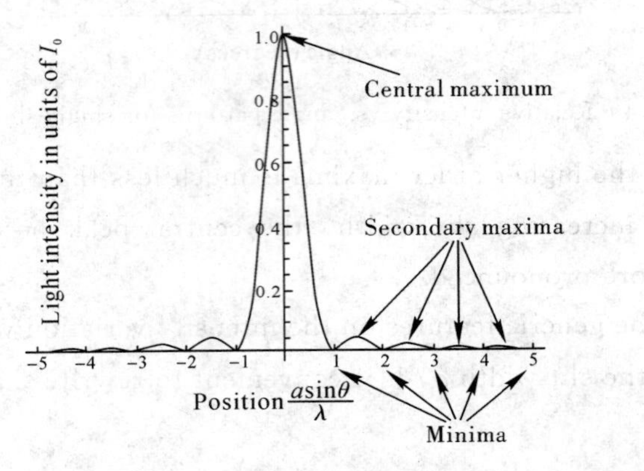


Fig. 12.5 Diffraction of light as a function of the ratio $\lambda/a$.

In Fig. 12.3, we can convert this into distance $y$ by using the approximation, $\sin\theta \approx \tan\theta = y/L$, where $L$ is the distance between the slit and the screen. Equation (12.2) becomes

$$a \cdot \frac{y}{L} = m\lambda \tag{12.7}$$

Thus, by measuring the distance between minima, you can determine the width single-slit.

2. Double-slit interference

For the double-slit case, we first analyze the idealized situation. Fig. 12.6 shows a coherent light source illuminating a double-slit diffraction grating. The slits have a width $b$ and separation $d$. We assume that a single light ray passes through the slit and illuminates a position on a distant screen. As before, we wish to know the intensity of the light at various positions on the screen. This is given by the expression:

$$I(\theta) = I_0 \cos^2\left(\frac{2\pi}{\lambda} d \cos\theta\right) \tag{12.8}$$

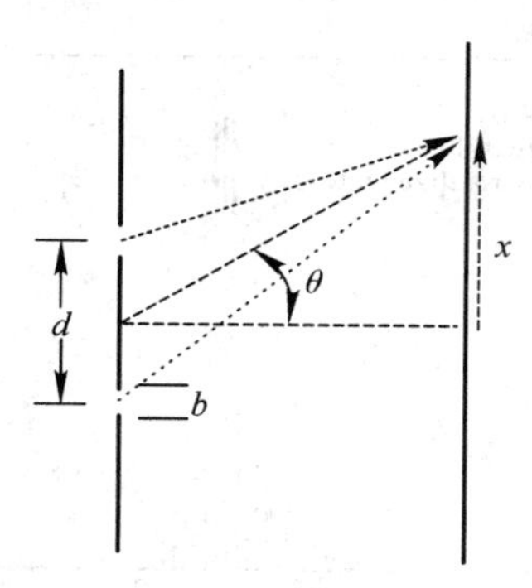


Fig. 12.6 Coherent light source illuminates a screen after passing through a double-slit diffraction grating. The slits have a width $b$ and are separated by a distance $d$. We wish to know the light intensity at a position $x$ on the screen.

Since the light from the two slits will have to travel different distances, they will either constructively or destructively interfere. The angular positions of the maxima are given by

$$d\sin\theta = m\lambda \tag{12.9}$$

The angular positions of the minima are given by

$$d\sin\theta = \left(m+\frac{1}{2}\right)\lambda \tag{12.10}$$

where $m=0, 1, 2, 3\cdots$ As in the single-slit case, we can approximate $\sin\theta \approx \tan\theta$ to get the spacings of maxima and minima in terms of $x$.

In this analysis, we have neglected the fact that the slits themselves have a finite width, and thus will exhibit single-slit diffraction. Therefore, in a real experiment like the one you are performing, the intensity on the screen will be a combination of both single-slit and double-slit diffraction. If the slit width $b$ is explicitly accounted for, then the intensity $I(\theta)$ is a convolution of the results obtained individually for the single- and double-slit cases. That is,

$$I(\theta) \propto \cos^2\left[\frac{2\pi}{\lambda}d\cos\theta\right]\left[\frac{\sin(\pi\, b\sin\theta/l)}{\pi b\theta/l}\right]^2 \tag{12.11}$$

With $d$ as the separation between slits and $b$ as the width of each, the normalized intensity is shown in Fig. 12.7.

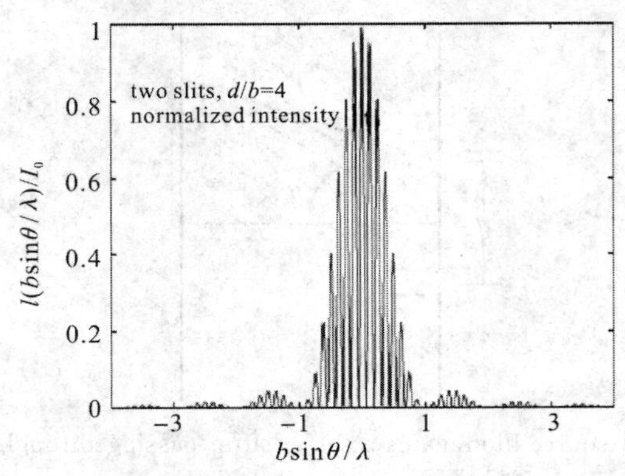


Fig. 12.7 Intensity plot considering both single-slit and double-slit diffraction.

The narrow maxima peaks are due to the double-slit interference, and the overall envelope pattern is due to single-slit diffraction. Thus, by measuring a diffraction pattern of a double-slit grating, you can determine the slit width and slit separation.

## Production

### Part 1 Diffraction and Single Slits

A laser produces monochromatic light ($\lambda = 632.8$ nm) which is well collimated and coherent. *Make sure your eyes are never exposed to direct laser light or its reflections.*

1. Align the laser beam with the slit.

① Mount the He-Ne laser at one end of the bench. Put the single-slit with holder on the optic bench a few centimeters away from the laser with the disk side of the holder closer to the laser. Plug in the laser and turn it on.

② Adjust the position of the laser beam from left-to-right and up-and-down until the beam is more or less centered on the slit. The screws to do this are on the back of the He-Ne laser. Once this position is set, it is not necessary to make any further adjustments of the laser beam when viewing any slits on the stand. When you move the stand to a new slit, the laser beam will already be aligned. The slits click into place so you can easily change from one slit to the next even in the dark.

2. Single slit diffraction

Set up a projection screen by clipping a piece of paper to the stand on the optical beam. Use the single slit wheel, which contains several single slits of different widths, in front of the laser so that a diffraction pattern is produced on the screen. Ensure that diffraction light is detected on a photocell.

Do the best compromise between autocollimation and centering you can (autocollimation attempts to make the slit wheel perpendicular to the beam). Record the patterns for the two available single slits. Then for each width, record the locations of the minima (≥ 3 on each side).

① From the pattern, the distance L from slit to screen and from $\lambda$, compute the slit width $a$ and estimate your error in $a$.

② Compare your measurement with the value indicated on the slit.

③ Observe and describe qualitatively the results of using the variable slit.

④ Extra Credit: Describe what you see for of the one of the two dimensional patterns or circular apertures. How is it related to the physical pattern on the disc?

**Part 2 Interference and Diffraction in Double Slits**

Switch holders to the double slits. Repeat procedures of Part 1 for this set of four double slits: quantitatively for one of the slits, and qualitatively for the rest.

From the observed pattern for the selected double slit, calculate both slit width and slit separation and their uncertainties for one of the double slits.

To do this you will measure the far-spaced minima for diffraction and the narrowly-spaced minima or maxima for interference. (You will need 5 or more mins or maxes on each side of 0)

For the other 3 double slits, compare the observed patterns with your expectations based on knowledge of "$b$" and "$d$". Describe the effects of the variable double slit, and use it to explain your results qualitatively.

## Questions

1. If you shine the laser beam on a hair, wire or line, why do you expect to see a diffraction pattern instead of just a shadow?

2. Why could there be a bright spot directly behind the obstacle? Think about the wire as being complementary to a slit.

# Lab 13　Construction of an Ammeter and Voltmeter

## Purpose

1. To determine the resistance of a galvanometer;

2. To study the design and construction of a DC ammeter with a full-scale reading of 100 mA and a DC voltmeter with a full-scale reading of 5 V.

## Apparatus

Galvanometer (with resistance $R_g$); microammeter; two decade resistance boxes; spool of resistance wire; multi-scale ammeter; multi-scale voltmeter; connecting wires; switch; DC power supply.

## Theory

Resistors are used in electric circuits to determine the current that flows through various branches of these circuits. They also determine the electric potential (commonly called voltage) at points in the circuit. Because of this, it is important to know the values of the resistance of each resistor in a circuit. The following well-known equation describes Ohm's law $U = IR$. This equation relates the resistance, $R$, the current through the resistor, $I$, and the voltage across the resistor, $U$.

The great majority of DC ammeters and voltmeters are based upon the simple galvanometer. A galvanometer is an instrument for detecting small currents. It consists of a coil of fine wire mounted so that it can rotate in the field of a permanent magnet. When current flows through the coil, the field of the permanent magnet exerts a torque on the coil and makes it rotate until equilibrium is established between the torque due to the field and that exerted by a restoring spring. Therefore, the angle of rotation depends on the current flowing through the coil and a needle attached to the coil can be calibrated to give the current. (In many designs, the field of the permanent magnet is uniform in the region of the coil and the angle is linearly proportional to the current, but for the galvanometer used in this lab, this is not quite the case.)

A typical analog ammeter consists of a sensitive galvanometer with a low resistance called a *shunt* connected in parallel with it. The shunt allows currents to flow through the ammeter which would otherwise burn out the galvanometer. The shunt also allows the ammeter to have a low resistance so that it will have a small effect on the circuit whose current is to be measured. (Recall that to measure current in a circuit, the circuit must be broken and the ammeter inserted between the break.)

A typical analog voltmeter consists of a sensitive galvanometer in series with a high resistance, known as a *multiplier*. Since a voltmeter is placed in parallel with the voltage to be measured, it should have a high resistance so that it does not appreciably alter the circuit being measured.

The range or full-scale reading of an ammeter or voltmeter can be changed by changing the shunt or multiplier resistor. From knowledge of the full-scale galvanometer current $I_g$, the galvanometer resistance $R_g$, and the desired range of the meter, the needed shunt or multiplier resistance can be easily calculated by means of Ohm's law.

## Procedure

### A. Determining $R_g$

To determine the internal resistance $R_g$, of the galvanometer, you will use the half scale method as described in Steps 1 - 4.

1. Turn off the power supply and turn the voltage output knob fully counter clockwise. Set both decade resistance boxes to their highest values. Connect the circuit as in Fig. 13.1. Connect in series the "galvanometer" (i. e. your microammeter), the DC power supply, and resistance box ($R_e$). Connect the switch $S_p$ and the other resistance box $R_p$ in parallel with the galvanometer ($R_g$). Open the switch $S_p$.

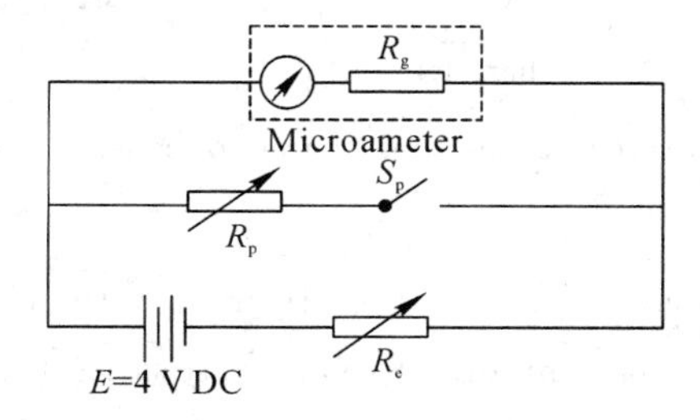


Fig. 13.1 The circuit connection diagram.

2. With $S_p$ open, turn on the power supply and slowly increase the supply voltage to 4 V. Gradually reduce $R_e$ until the microammeter reads full scale, $I_g = 0.5$ mA. Record $R_e$.

3. Close the switch $S_p$ and then gradually reduce $R_p$ until the microammeter reads exactly one-half of full scale, $I_g/2 = 0.25$ mA. Record $R_p$.

4. Using your values of $R_e$ and $R_p$, calculate $R_g$ from

$$R_g = \frac{R_e R_p}{R_e - R_p}$$

A relation is derived in the Appendix. If your value of $R_g$ is not in the range 300 - 500 Ω, repeat Steps 1 - 4.

**B. Constructing an Ammeter with a Full Scale Reading of 100 mA.**

1. Preliminary Work

(1) Study Fig. 13.2, and calculate the value of the shunt resistance $R_s$ which will be required to make an ammeter with a full-scale reading of 100 mA. Note that the same voltage is across both the shunt and $R_g$.

(2) Cut the appropriate length of resistance wire to create a shunt of that resistance, plus a little extra to connect at the binding posts. The instructor will give you the linear resistivity (ohms per foot or per cm) of the wire to be used.

2. Connecting the Circuit

Turn off the power supply. Wire the circuit shown in Fig. 13.2. Set one decade box to 40Ω and use it as the load resistor $R_L$.

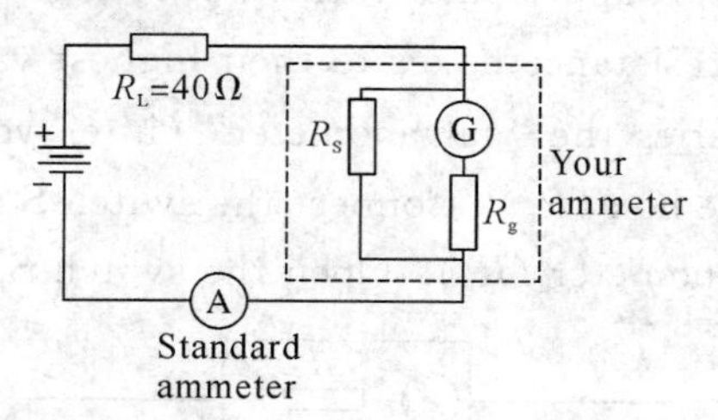


Fig. 13.2 The circuit connection diagram of constructing an ammeter.

3. Calibrating Your Ammeter

(1) Slowly vary the power supply voltage from zero to several volts to find the range of voltage which will cause your ammeter to vary across its full range from 0 to 100 mA. Starting at 0 V, choose at least six voltage settings spanning this range and record the corresponding current reading on your ammeter and the standard ammeter.

(2) Plot the current readings of the standard ammeter vs the corresponding readings on your ammeter. Connect the points with a smooth line. This curve is a calibration curve and would be a straight line with slope of 1 if your ammeter agreed exactly with the standard ammeter. Discuss how this curve could be used to improve the accuracy of the current readings of your ammeter.

### C. Constructing a Voltmeter with a Full Scale Reading of 5 V.

Use Ohm's law to calculate the value of the multiplier resistor $R_m$ that must be in series with your galvanometer resistance $R_g$ to give full-scale deflection $I_g$ when the voltage applied to the series combination is 5 V.

### D. Wiring and Calibrating your Voltmeter

(1) Set the calculated multiplier resistance $R_m$ on a decade resistance box, connect it in series with the galvanometer, and connect the digital voltmeter across the combination as shown in Fig. 13. 3.

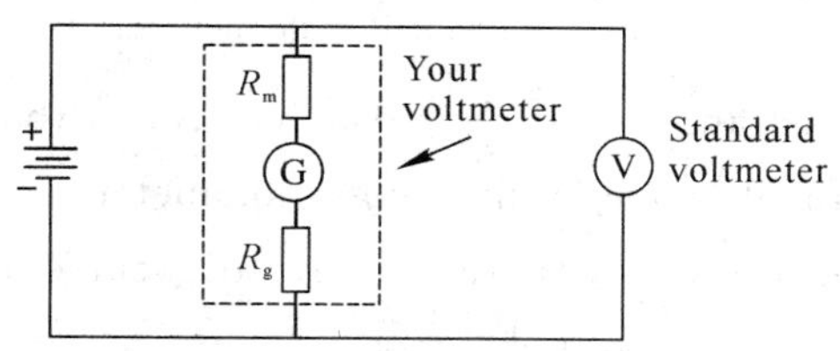


Fig. 13. 3 The circuit connection diagram of constructing a voltmeter.

(2) Using at least six voltage settings between 0 V and 5 V, measure and record the readings of your voltmeter and of the digital voltmeter. Make a calibration curve by plotting these readings of the digital voltmeter vs the readings of your voltmeter. Connect the points on the graph by a smooth curve. The result should be a straight line.

## Questions

1. What resistance would an ideal ammeter have? (Why?)
2. What resistance would an ideal voltmeter have? (Why?)
3. If you had an ammeter with a maximum current rating of 1 A and an internal resistance of 100Ω, what shunt resistance would allow you to increase its range to 100 A?

## Appendix—Half Scale Method

The half scale method is a very accurate way for determining the resistance of a galvanometer.

1. First, set up the circuit shown in Fig. 13.4, leaving the switch open. For this configuration the total current in the circuit $I$ goes through the galvanometer. Adjusting this current to maximum reading gives

$$U = I_g R_g + I_g R_e = I_g (R_g + R_e)$$

This defines the maximum current through the galvanometer $I_g$.

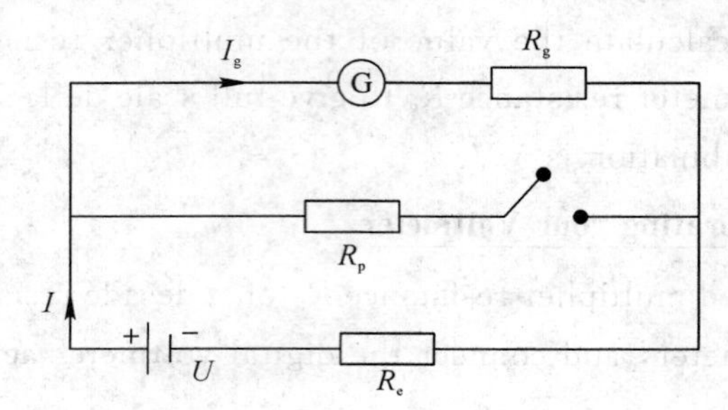


Fig. 13.4 The half scale method.

2. Now close the switch and adjust the resistance $R_p$ so that the current through the galvanometer is $1/2\ I_g$ defined above (thus the galvanometer reads 1/2 maximum).

Since the voltage is left constant, there is a new total current in the circuit $I = 1/2\ I_g + I_p$. Since the meter and $R_p$ are in parallel

$$\frac{1}{2} I_g R_g = I_p R_p \quad I_p = \frac{1}{2} I_g \frac{R_g}{R_p}$$

Substituting this value of $I_p$ into the expression for the total current:

$$I = \frac{1}{2} I_g \left( \frac{R_p + R_g}{R_p} \right)$$

We can rewrite the voltage $U$ in terms of $R$, the parallel combination of $R_g$ and $R_p$. That is $U = I(R + R_e)$ with

$$R = \frac{R_p R_g}{R_p + R_g}$$

Equating the expressions for $U$ before and after closing the switch then gives

$$I(R + R_e) = I_g (R_g + R_e)$$

Making the various substitutions

$$\frac{1}{2} I_g \left( \frac{R_p + R_g}{R_p} \right) \left( \frac{R_p R_g}{R_p + R_g} + R_e \right) = I_g (R_g + R_e)$$

Dividing out $I_g$ and rewriting this equation we find

$$\left(\frac{R_p+R_g}{R_p}\right)\left(\frac{R_pR_g+R_eR_p+R_eR_g}{R_p+R_g}\right)=2(R_g+R_e)$$

The term $R_p + R_g$ cancels, leaving $R_pR_g + R_eR_p + R_eR_g = 2R_gR_e$. Bringing terms involving $R_g$ to the left hand side and all others to the right, and summing the terms gives $R_g(R_e - R_p) = R_eR_g$ and thus the resistance of the galvanometer is found:

$$R_g=\frac{R_eR_g}{R_e-R_p}$$

# Lab 14 Electric and Magnetic Forces

## Purpose

To study the deflection of a beam of electrons by electric and magnetic fields.

## Apparatus

Electron beam tube; stand with coils; power supply; multimeters.

## Theory

This experiment uses a large vacuum tube which contains an "electron gun". The gun is a metal filament that is heated so hot that some of the electrons of the metal have enough kinetic energy to leave the filament. They are then accelerated by a potential difference $V_a$ which gives them a kinetic energy $mv_x^2/2$, equal to $eV_a$. ($e$ is charge of an electron, $-1.6\times 10^{-19}$ coulombs, and $m$ is its mass, $9.1\times 10^{-31}$ kgm.) The electron beam becomes visible when the electrons strike a fluorescent screen.

As the electrons move in the horizontal ($x$) direction, an electric force may be applied in the vertical ($y$) direction by applying a potential difference $V_y$ between two deflecting plates. The apparatus also permits vertical deflection by a magnetic force produced by a "Helmholtz coil".

In the first part of the experiment we will study just the magnetic deflection; in the second part only the electric deflection; and, in the third part a combination of both forces applied in opposite directions so as to cancel each other's effect.

1. Magnetic field deflection

In this part of the experiment, the speed of the electron, $v_x$, will be determined by measuring the effect of a magnetic field on the path of the electrons. The magnetic field ($B$) of the coils is horizontal, and at right angles to the direction of the electron beam. It produces a force on the electrons whose magnitude is $F_m = ev_xB$.

The magnetic force is at right angles to the electron velocity; thus, the direction of motion of the electrons will be changed but their speed will remain constant. The electrons will move in a circular path of radius $R$ with the force toward the center of motion being the magnetic force. The relation between the radial force and the magnetic force is

$$ev_x B = \frac{mv_x^2}{R} \quad \text{or} \quad v_x = \frac{BRe}{m} \tag{14.1}$$

The magnetic field is produced by a pair of current-carrying coils. The field $B$ produced by the coils is proportional to the current $I$ flowing in the coils. For the coils used for this experiment use the relationship, $B = 4.23 \times 10^{-3} I$, where $B$ is in Teslas and $I$ is in amps. $R$ is measured on the screen. Fig. 14.1 shows that, for the deflected beam, $x$, $y$, and $R$ are related by $R^2 = (R-y)^2 + x^2$, so that

$$R^2 = \frac{x^2 + y^2}{2y} \tag{14.2}$$

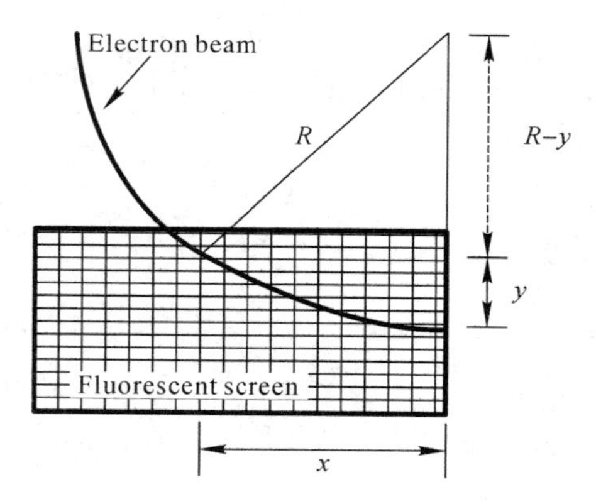


Fig. 14.1 The electron beam travels to the left and becomes visible on the fluorescent screen.

2. Electric field deflection

In this part of the experiment, $y$ deflection is produced by an electric field $E_y$. The vertical electric force is $eE_y$ and the vertical acceleration is $eE_y/m$. The force acts for a time $t$ so that (with $v_{y-0} = 0$)

$$y = \frac{1}{2} a_y t^2 - \frac{1}{2} \frac{eE_y}{m} t^2$$

The $x$-component of the velocity of the electrons is constant at a value $v_x$, with $x = v_x t$, so that we can write

$$y = \frac{1}{2} \frac{eE_y x^2}{mv_x^2}$$

In this relation, all quantities except $x$ and $y$ are constant, so that the relation has the form $y=nx^2$, where $n$ is a constant. This is the equation of a parabola.

The value of $v_x$ is given by conservation of energy (electric potential energy=kinetic energy)

$$eV_a=\frac{1}{2}mv_x^2$$

So we can substitute $2eV_a$ for $mv_x^2$

$$y=\frac{E_y}{4V_a}x^2$$

If the deflection plates were very large compared to their separation $d$, then $E_y$ would be given by $V_y/d$. Since this is not the case, $E_y$ will be different by a correction factor $k$, so that

$$E_y=k\frac{V_x}{d} \tag{14.3}$$

Because we use the same voltage for the acceleration ($V_a$) as for the deflection ($V_y$) ($V_a=V_y$) we can therefore write

$$y=k\frac{V_y}{4dV_a}x^2$$

and finally,

$$y=k\frac{1}{4d}x^2 \tag{14.4}$$

## Procedure

1. Electric deflection

(1) Turn on the power supply.

(2) Adjustaccelerated voltage $V_2$ and focus voltage $V_1$ to make the light on the tube screen to the dot. Attention: The light dot is not too bright.

(3) Put "high voltage measurement" in theaccelerated voltage file $V_2$, record the accelerated voltage.

(4) Put "voltage and current measurement" in "$V_{dx}$" and "$V_{dy}$", respectively, make the light dot lie in the center by adjusting "$V_{dx}$" "$V_{dy}$" and "$X$, $Y$ zero potentiometer".

(5) Put "voltage and current measurement" in "$V_{dx}$", and vary accelerated voltage $V_2$. Observe the deflected beam and record the values of D for $V_{dx}$.

(6) Plot $D$ against $V_{dx}$ with the maximum and minimum element in the array to the same coordinate system, and determine the slope from the graph and calculate the electric deflection sensitivity $\varepsilon_x = \Delta D / \Delta V_{dx}$.

(7) Adjust the "$V_{dx}$" to zero, put "voltage and current measurement" in "$V_{dy}$", and repeat steps (5) and (6).

**Table 14.1 Electric deflection sensitivity along the *x*-axis.** voltage unit: V

| $D$/mm | | | | −20 | −15 | −10 | −5 | 0 | 5 | 10 | 15 | 20 |
|---|---|---|---|---|---|---|---|---|---|---|---|---|
| 1 | $V_2$ | | $V_{dx}$ | | | | | | | | | |
| 2 | $V_2$ | | $V_{dx}$ | | | | | | | | | |
| 3 | $V_2$ | | $V_{dx}$ | | | | | | | | | |

**Table 14.2 Electric deflection sensitivity along the *y*-axis.** voltage unit: V

| $D$/mm | | | | −20 | −15 | −10 | −5 | 0 | 5 | 10 | 15 | 20 |
|---|---|---|---|---|---|---|---|---|---|---|---|---|
| 1 | $V_2$ | | $V_{dx}$ | | | | | | | | | |
| 2 | $V_2$ | | $V_{dx}$ | | | | | | | | | |
| 3 | $V_2$ | | $V_{dx}$ | | | | | | | | | |

2. Magnetism deflection

(1) Turn on "constant current source" switch.

(2) Put "voltage and current measurement" in "200 mA" files.

(3) Adjust the potentiometer of "constant current source adjustment" fully counterclockwise to make the current zero, vary the "$X$, $Y$ zero potentiometer" to make the light dot lie in the center.

(4) Adjust the potentiometer of "constant current source adjustment" clockwise, record the data about current "$I_m$" and offset distance "$D$".

(5) Change the "reversing switch" at the mid left of the oscillograph tube screen; repeat Steps (3) and (4).

(6) Plot $D$ against $I_m$ with the maximum and minimum element $V_2$ in the array to the same coordinate system, and determine the slope from the graph and calculate the magnetism deflection sensitivity $\delta$.

**Table 14. 3 Magnetism deflection sensitivity along the $y$ axis.** current unit: mA

| $D$/mm | | | | −20 | −15 | −10 | −5 | 0 | 5 | 10 | 15 | 20 |
|---|---|---|---|---|---|---|---|---|---|---|---|---|
| 1 | $V_2$ | | $I_m$ | | | | | | | | | |
| 2 | $V_2$ | | $I_m$ | | | | | | | | | |
| 3 | $V_2$ | | $I_m$ | | | | | | | | | |

# Lab 15 Velocity of Sound in Air

## Purpose

1. To measure the wavelength, frequency, and propagation speed of ultrasonic sound waves;

2. To observe interference phenomena with ultrasonic sound waves;

3. To understand the piezoelectric effect;

4. To study the methods to measure the speed of sound waves.

## Apparatus

Oscilloscope: function generator; ultrasonic transducers; meter stick; angle board; a vernier calipers; thermometer; coaxial cables.

## Theory

The transmission of sound can be illustrated by using a model consisting of an array of balls interconnected by springs. For real material the balls represent molecules and the springs represent the bonds between them. Sound passes through the model by compressing and expanding the springs, transmitting energy to neighboring balls, which transmit energy to their springs, and so on. The speed of sound through the model depends on the stiffness of the springs (stiffer springs transmit energy more quickly). Effects like dispersion and reflection can also be understood using this model.

In a real material, the stiffness of the springs is called the elastic modulus, and the mass corresponds to the density. All other things being equal (ceteris paribus), sound will travel more slowly in spongy materials, and faster in stiffer ones. For instance, sound will travel 1.59 times faster in nickel than in bronze, due to the greater stiffness of nickel at about the same density. Similarly, sound travels about 1.41 times faster in light hydrogen gas than in heavy hydrogen (deuterium) gas, since deuterium has similar properties but twice the density. At the same time, "compression-type" sound will travel faster in solids than in liquids, and faster in liquids than in gases, because the solids are more difficult to compress than liquids, while liquids in turn are more difficult to compress than gases.

The motion of the elements of the medium in a longitudinal wave is back and forth along the direction in which the wave travels. Fig. 15.1 illustrates the technique used to produce ultrasonic waves for clinical use. Electrical contacts are made to the opposite faces of a crystal, such as quartz or strontium titanate. If an alternating voltage of high frequency is applied to these contacts, the crystal vibrates at the same frequency as the applied voltage, emitting a beam of ultrasonic wave. At one time, this was how almost all headphones produced sound. This method of transforming electrical energy into mechanical energy, called the piezoelectric effect, is also reversible. If some external source causes the crystal to vibrate, an alternation voltage both generates and receives ultrasonic waves.

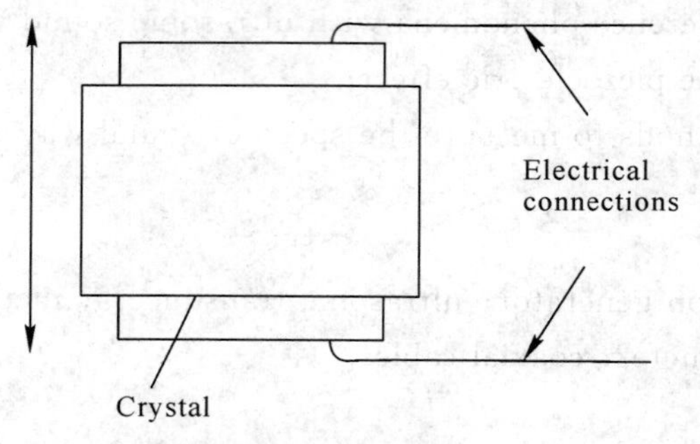


Fig. 15.1 An alternating voltage applied to the faces of a piezoelectric crystal causes the crystal to vibrate.

We know that the speed of sound also depends on the temperature of the medium. For sound traveling through air, the relationship between the speed of sound and temperature is

$$V=V_0\sqrt{\frac{T}{T_0}}=V_0\sqrt{1+\frac{t}{273.15}} \tag{15.1}$$

where $V_0$ is the speed of sound in air at 0℃, $T$ is the absolute (Kelvin) temperature and $t$ is the Celsius temperature (℃). Using this equation, we find that at 293 K, the speed of sound in air is approximately 343 m/s.

The wave speed $V$ can be expressed as follows:

$$V=\frac{\lambda}{f} \tag{15.2}$$

where $\lambda$ is wavelength and $f$ is frequency of wave.

In this experiment, energy of ultrasonic wave can be converted from electrical energy by transducers constricted of piezoelectric materials when the frequency of ultrasonic wave is the same as the transducers frequency. In other words, the frequency of ultrasonic wave $f$ can be found when intensity of ultrasonic wave stimulated by alternating current source with the same frequency reaches the maximum.

Then, there are two main means to measure the wavelength $\lambda$ of ultrasonic wave. One is based on the property of standing wave while the other is based on phase comparison between the waves. The circuit configuration of measuring the speed of sound with oscilloscope is shown in Fig. 15. 2.

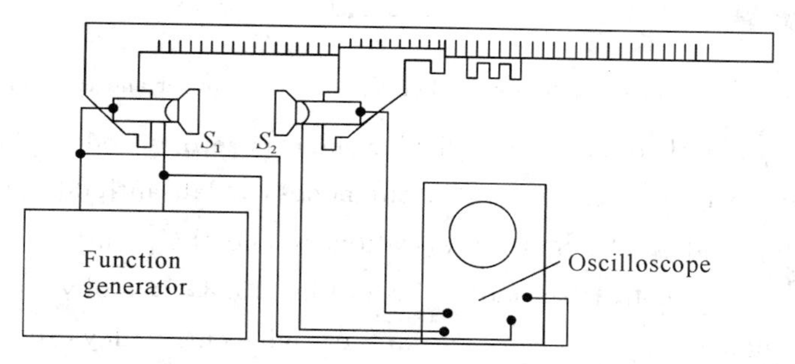


Fig. 15. 2 The circuit configuration of measuring the speed of sound with oscilloscope.

1. Based on the property of standing wave

The distance between a pair of transducers, in which the left one is transmitter and the right one is receiver, is $L$. The function $Y_1$ of the wave stimulated from the transmitter can be written as

$$Y_1 = A\cos\left(\omega t - \frac{2\pi}{\lambda}x\right) \tag{15. 3}$$

where $A$ is intensity, $\omega$ is angular frequency of wave, and $\lambda$ is wavelength.

The receiver (right one) not only receives, but also reflects part of the coming ultrasonic wave from the transmitter. Considering half-wave loss, the function $Y_2$ of the reflective wave which has the same magnitude of displacement but opposite sign can be written as

$$Y_2 = A\cos\left(\omega t + \frac{2\pi}{\lambda}x + \pi\right) \tag{15. 4}$$

Then, the standing wave can be produced by interfering the traveling wave from transmitter and reflective wave from receiver. Its wave functions can be expressed as

$$\begin{aligned} Y &= Y_1 + Y_2 \\ &= A\cos\left(\omega t - \frac{2\pi}{\lambda}x\right) + Y_1 = A\cos\left(\omega t + \frac{2\pi}{\lambda}x + \pi\right) \end{aligned}$$

$$=2A\sin(\omega t)\sin\left(\frac{2\pi}{\lambda}x\right) \quad (15.5)$$

This is the function of standing wave, shown in Fig. 15.3.

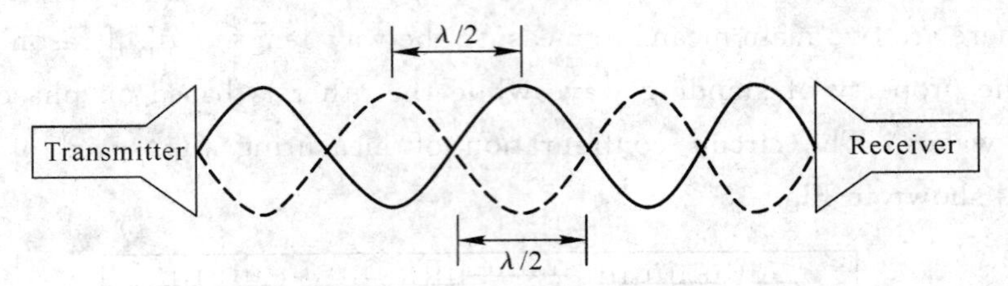


Fig. 15.3 The standing wave between the transmitter and receiver.

According the wave theory, the net displacement is zero at nodes. There is no motion at the nodes but midway between two adjacent nodes, at an antipode, the wave vibrates with the largest amplitude. The nodes are stationary and the points at which the wave is attached to the transducers are also nodes. From Fig. 15.3, it is obvious that the distance between adjacent nodes, or adjacent peaks, or adjacent valleys is one half of the wavelength of the wave:

$$d=\frac{\lambda}{2} \quad (15.6)$$

So, once the distance between adjacent nodes, or adjacent peak, or adjacent valley is measured, the wavelength of the ultrasonic wave $\lambda$ can be obtained by Equation (15.6).

2. Based on the phase comparison

The function Y of the wave from the transmitter can be written as

$$Y_1=A\cos\left(\omega t-\frac{2\pi}{\lambda}x\right) \quad (15.7)$$

At the point of the transmitter, the function of the wave is

$$Y_1=A\cos\left(\omega t-\frac{2\pi}{\lambda}x\right)\bigg|_{x=0}=A\cos\omega t \quad (15.8)$$

Comparing the functions of the waves at the points of transmitter and receiver, we can find that there is a phase difference $\Delta\varphi$ between them, $\Delta\varphi=2\pi\Delta L/\lambda$. In other words, the phase difference $\Delta\varphi$ will change $2\pi$ when the distance $\Delta L$ between transmitter and receiver is changed from $L_0$ to $L_0+\lambda$, where $L_0$ is the initial distance between transmitter and receiver. We can measure the distance corresponding to one wavelength through observing Lissajous patterns which are the superposition of two sinusoidal waves with the same frequency from transmitter and receiver. In one period, the Lissajous patterns (Fig. 15.4) which include line with positive slope, ellipse, circle, ellipse, line with negative slope, ellipse, circle, ellipse, should appear in turn. In order to identify one period easily, it is best to choose one line pattern (positive slope or negative slope) as the beginning pattern.

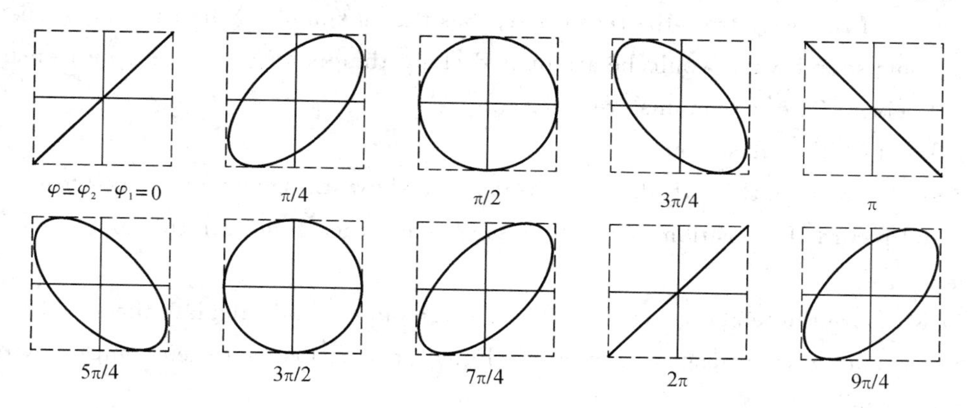


Fig. 15.4 Lissajous patterns observed.

When the same patterns appear one after another, the phase difference $\Delta\varphi$ is $2\pi$ and the distance difference $\Delta L$ between transmitter and receiver is $\lambda$.

## Procedure

(1) Connect and adjust oscilloscope. Connect the circuit shown in Fig. 15.2, and adjust the distance $L$ between the transmitter and the receiver about 5 – 6 cm. Check it again. Turn on the oscilloscope and the function generator. Choose sinusoidal wave with frequency 35.000 kHz from function generator. Adjust the oscilloscope so that the sinusoidal wave is clear and centered on the screen of oscilloscope.

(2) Search for the frequency $f$ of the ultrasonic sound wave. Change the frequency of sinusoidal wave from the transmitter and the receiver at one time. You will find that the wave from the receiver changes both with frequency and intensity while the wave from transmitter changes only with frequency. When the intensity of the wave from the receiver reaches the maximum, the frequency of the ultrasonic sound wave, which is the same as the frequency of sinusoidal wave from the function generator, is found. Record the frequency and keep the frequency in the whole experiment.

(3) Measure the wavelength $\lambda$ of the ultrasonic sound wave and calculate the speed $v$ of the ultrasonic sound wave by Equation (15.2).

① Based on the property of standing wave

Increase the distance between the transmitter and the receiver slowly. Observe the wave from the receiver through oscilloscope and record the positions $l_i$ in Table 15. 1 when the intensity of the wave from the receiver reaches the maximum. Note that the intensity of the ultrasonic sound wave would be attenuated along the distance between the transmitter and the receiver. The maximums are different.

② Based on the phase comparisons

Press the X-Y button of the oscilloscope to show Lissajous pattern. Observe the pattern and record the position $l_i$ in Table 15. 2 when the same patterns appear one after another.

(4) Measure the temperature of air using thermometer and calculate the speed $v$ of the ultrasonic sound wave propagation in air by Equation (15. 1), and then compare with the measured value.

## Record and Calculation

**Table 15. 1 Measurement of wavelength by property of standing wave.** $f=$ kHz

| $i$ | $l_i$/mm | $\Delta l_i = l_{i+6} - l_i$/mm |
|---|---|---|
| 1 | | |
| 2 | | |
| 3 | | |
| 4 | | |
| 5 | | |
| 6 | | |
| 7 | | |
| 8 | | |
| 9 | | |
| 10 | | |
| 11 | | |
| 12 | | |

$\Delta\bar{L} = \Delta\bar{l}/n =$ ____, $\lambda = 2\Delta\bar{L} =$ ____. The speed of the ultrasonic sound wave $v = \lambda \cdot f =$ ____

**Table 15.2 Measurement of wavelength by phase comparison.** $f=$ kHz

| $i$ | $l_i$/mm | $\Delta l_i = l_{i+6} - l_i$/mm |
|---|---|---|
| 1 | | |
| 2 | | |
| 3 | | |
| 4 | | |
| 5 | | |
| 6 | | |
| 7 | | |
| 8 | | |
| 9 | | |
| 10 | | |
| 11 | | |
| 12 | | |

$\Delta \bar{L} = \Delta \bar{l}/n =$____, $\lambda = 2\Delta \bar{L} =$____. The speed of the ultrasonic sound wave $v = \lambda \cdot f =$____

## Questions

1. Is the speed of sound independent of frequency? Does the speed of sound depend on anything that could be easily varied in this experiment?

2. Compare and contrast transverse and longitudinal waves.

3. What limited the number of resonances you found in the tube in this experiment for a given frequency?

4. In order to find the Eigen frequency $f$ of transducers, is it right to measure through oscilloscope by adding the traveling wave from the transmitter and the deflective wave from the receiver? Why?

# Lab 16  The Ballistic Galvanometer

## Purpose

1. To investigate characteristics of galvanometer;
2. To understood galvanometers and how they can be used to measure small currents;
3. To understood the accuracy and precision to which current can be measured.

## Apparatus

The galvanometer; voltmeter; rheostat resistor; standard resistance; resistance box; switch; wire.

## Theory

A galvanometer is a very sensitive coil meter. If there is a small voltage across the terminals of the galvanometer, there will be a deflection of the needle. The direction of the needle's deflection indicates the polarity of the voltage. If there is a small current through the galvanometer, there will also be a deflection indicating the direction as shown in Fig. 16. 1. Basically, there is just a coil of wire connected between the two ports on the galvanometer. The coil is in a magnetic field, so it will tilt when there is a current through it, or a voltage across it.

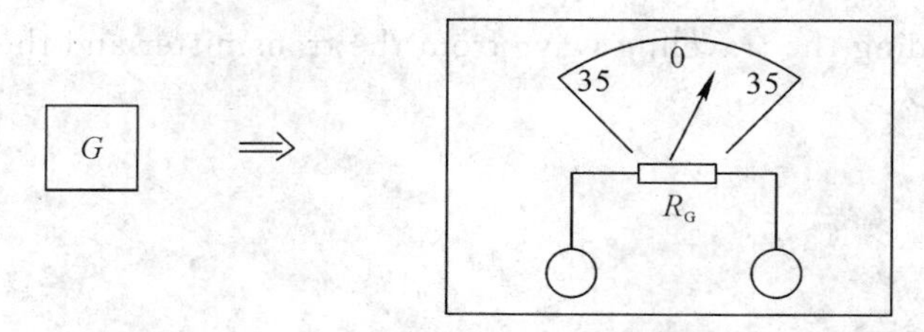


Fig. 16. 1  Model of inner circuit of galvanometer.

There is one limitation to the use of a galvanometer—it is very sensitive to voltages and currents—too sensitive for most measurements. But, we can use this simple coil meter to create ammeters and voltmeters with any desired range.

In Fig. 16.2 if $(P+G)$ is much larger than $R$, then the total resistance of the galvanometer part can be approximated to $R$. The resistors $Q$ and $R$ form a potential divider. If $Q$ is much larger than $R$, the potential difference, $V_R$ is small. Write down an expression that gives $V_R$ in terms of $E$, $Q$ and $R$ and evaluate the resistance $Q$ required to produce a voltage drop $V_R$ of $5\times10^{-4}$ V assuming that $E=5$ V. [Hint: The resistance of the galvanometer is given by $\frac{1}{R_{total}}=\frac{1}{R}+\frac{1}{P+G}$.

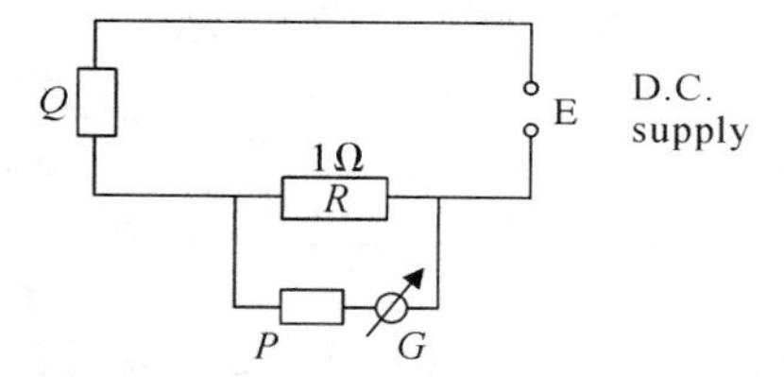


Fig. 16.2 Reinforce the series/parallel circuit.

The resistance $R$ and the combination of $P$ and $G$ in parallel with $R$ act as a current divider, limiting the current flowing through the galvanometer.

By explicitly writing Ohm's law for each leg of the current divider, and using the expression for the voltage, $V_R$ found above, it shows that the current flowing through the combination of $P$ and $G$ is given by:

$$I_{(P+G)}=\frac{V_R}{P+G}=\frac{ER}{(R+Q)(P+G)} \tag{16.1}$$

If you look closely at an analogue ammeter or voltmeter you will see that there the deflection of the arm to give the reading is controlled through a magnetic coil.

As shown in Fig 16.3, a galvanometer consists of a coil of wire wound around a plastic former. This is suspended between the poles of a magnet. When current passes to the galvanometer coil, it produces a magnetic field which interacts with the field produced by the permanent magnet. This *torque*, causes the coil to rotate, and a pointer fixed to the coil to move. The galvanometer you will be using has a mirror attached to the coil, and light is reflected from the mirror onto the scale on the front of the machine.

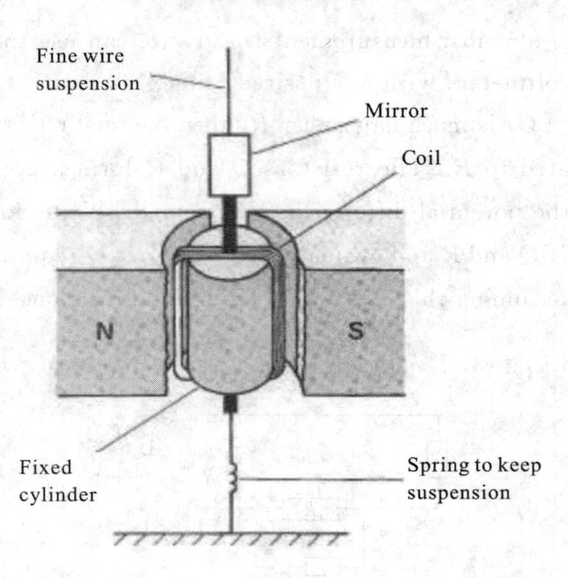


Fig. 16. 3 Structure of the galvanometer.

In Fig. 16. 2, identify the D. C. power supply $E$, the hand-held multimeter and the 1 Ω resistor, $R$. Identify the resistance boxes. By turning the dials on the resistance boxes, the resistance can be varied from a few ohms up to several thousand ohms.

The resistance of the galvanometer is governed by the metallic coil. Typical values of resistance of metallic coils are in the range of 1 to 1000 Ω. This range is controlled by the composition and amount of the coil material. Therefore, in order to use a galvanometer as a sensitive voltmeter we will need to add a large resistance in series with it. Recall from skills session 1 that the resistance of a voltmeter should be very high.

**1. Finding the internal resistance of the galvanometer**

One of the principal reasons for using a galvanometer is their sensitivity to very small currents. We will therefore need to limit the voltage that will be dropped across the galvanometer, and hence the current that flows through it. To achieve this we construct potential and current dividers.

In Fig. 16. 4, the galvanometer with the resistance box $P$ in series with it act as a voltmeter, measuring the voltage drop across the 1 Ω resistor, $V_R$. The current that flows through the galvanometer part of the circuit is supplied by this potential difference. $V_R$ is made small by dropping the source voltage $E$ (approx 5 V) across the potential divider $Q$

and R, with $Q$ much larger than $R$.

The current flowing through the combination of $P$ and $G$ will induce a torque on the galvanometer coil which in turn will deflect the light beam a distance $d$, the magnitude of $d$ being proportional to the current flow:

$$d=k_i I \tag{16.2}$$

The constant of proportionality $k_i$ is defined as the current sensitivity, and so from Equation (16.1) we can obtain

$$d=k_i \frac{E}{(P+G)} \frac{R}{(R+Q)} \tag{16.3}$$

In this experimental configuration, setting $P=1,000\ \Omega$ and $Q=10,000\ \Omega$ should produce a current flow of approximately $5\times10^{-7}$ A through the galvanometer which will produce a torque displacing the beam on the galvanometer by a couple of centimetres.

Equation (16.3) can be re-arranged into the following forms:

$$\frac{RE}{R+Q}\frac{K}{d}=P+G \quad \text{or} \quad P=\frac{RE}{R+Q}\frac{K}{d}-G \tag{16.4}$$

The values of $K$ and $G$ can be found by varying $P$ and recording the displacement of the galvanometer beam.

Set the power supply to give 5 V. Record its value and uncertainty in your lab book. Construct the circuit below using the components you have already identified.

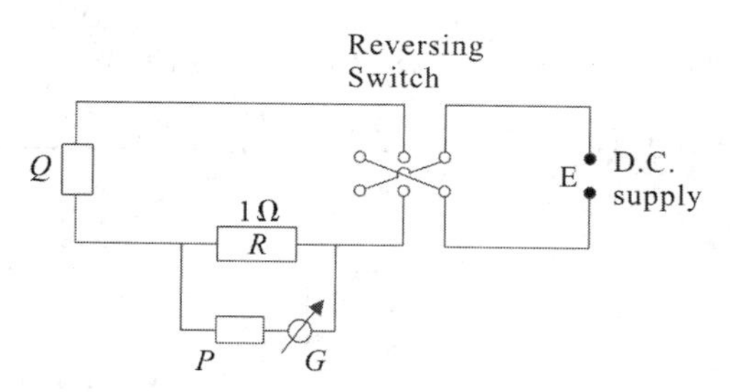


Fig. 16.4 Measurements the internal resistance of the galvanometer.

Initially set $P=1000\ \Omega$ and $Q=10,000\ \Omega$. This should cause the beam to move by a couple of centimetres. Reduce $P$ and increase $Q$ to keep the deflection on the scale. Continue until $P$ is zero and then adjust $Q$ to give an almost full scale deflection. Then keep $Q$ fixed and make a note of its value and uncertainty in your lab book.

Now make a series of measurements of the deflection $d$ as $P$ is increased.

From Equation (16.4), a plot of $P$ against $1/d$ will give a straight line with an intercept of $G$. However, as it is $P$ that we are varying (the dependent variable), this should be plotted on the $x$ axis. The straight line graph you should plot is $\beta/d = P/k + G/k$ (c. f. $y = mx + c$) where $\beta = R \cdot E/(R+Q)$. Calculate $\beta$ and its error, and use a spreadsheet to generate the data in a form that will be suitable for plotting.

[Hint: The % errors on $R$, $E$, $Q$ are all approximately 1%. The error on $R+Q$ is the same as that of $Q$ alone. Thus, we can approximate the error on $\beta$ as being 3%.]

**2. The torsion pendulum and the ballistic galvanometer**

The ballistic galvanometer is a galvanometer adopted to measure charge as opposed to current. It measures the charge flowing through the system during the passage of a transient current and is based on the ideas behind the torsion pendulum.

If the period of oscillation is long in comparison with the transient current, the electromagnetic pulse created in the coil acts as the impulse to start the coil oscillating. The angular displacement, $\theta$ is directly related to the charge, $Q$. The galvanometer you are using has been selected because the coil has a high moment of inertia which results in a long oscillation period.

However, because of effects of air-resistance and other 'resistive' effects within the coil, these oscillations will be damped, and the coil will eventually come to rest at the equilibrium point.

Rebuild your circuit so that the following is realized (See Fig. 16.5):

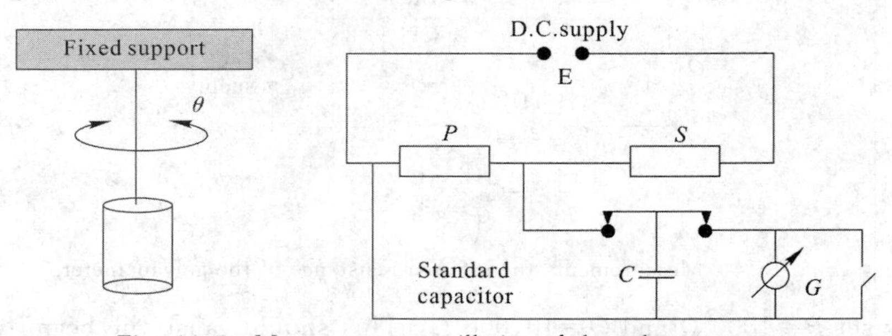


Fig. 16.5 Measurements oscillation of the galvanometer.

This time the potential dividers $P$ and $S$ provide a voltage to charge a standard capacitor, $C$. This capacitor can be discharged into the galvanometer using the Morse key.

The amount of charge, $Q$, discharged is simply given by $Q = VC$ with the voltage dropped across the capacitor, $V$, given by the potential dividers $P$ and $S$. Thus we have that the charge is given by

$$Q = EC\frac{P}{P+S} \tag{16.5}$$

This instantaneous current (charge pulse) provides the impulse to set the galvanometer into oscillations. The coil begins to swing, but because of damping it will not reach its maximum deflection $a_0$. Instead it is deflected to $a_1$ on the first swing, $a_2$ on the second, and so on, at the following times:

$$\begin{aligned} t &= \frac{T}{4} \quad \frac{3T}{4} \quad \frac{5T}{4} \\ d &= a_1 \quad a_2 \quad a_3 \end{aligned} \tag{16.6}$$

The amplitude of the swings is being damped by an exponential which, as a function of time, $t$, has the following form:

$$A = \exp(-\gamma t) \tag{16.7}$$

with $\gamma$ the damping term. It follows then that the ratio of successive amplitudes is related to the period of oscillation, $T$, and is given by:

$$\frac{a_1}{a_2} = \frac{a_2}{a_3} = \frac{a_3}{a_4} \cdots = \exp\left(\frac{\gamma T}{2}\right) \tag{16.8}$$

where $\gamma = (BAn)^2/(2IX)$. $B$ is the flux density of the magnetic field in which the coil sits, $A$ is the area of the coil and $n$ is the number of turns of wire in the coil. $T$ is the period of the oscillations and $X$ the total resistance of the galvanometer circuit.

The maximum deflection, which would be achieved in the absence of any damping, is given by $a_0$. It is found as follows (using Equation (16.8)):

$$\frac{a_0}{a_1} = \exp\left(\frac{\gamma T}{4}\right) = \left[\exp\left(\frac{\gamma T}{2}\right)\right]^{\frac{1}{2}} = \sqrt{\frac{a_1}{a_2}} \qquad a_0 = a_1\sqrt{\frac{a_1}{a_2}} \tag{16.9}$$

The charge sensitivity can be defined as the maximum amplitude divided by the charge. Therefore from Equation (16.5) we can write

$$K = \frac{a_0}{Q} = a_0\frac{P+S}{ECP} \quad \text{or} \quad a_0 = K\frac{ECP}{P+S} \tag{16.10}$$

**3. The period and relating *k* and *K***

Make $(P+S)$ a simple multiple of $E$ (say $1000 \times E$) and keep it constant. Set the capacitor to be 0.5 μF. For a range of values of $P$ (keeping $P+S$ fixed) measure the deflections $a_1$ and $a_2$. Use an Excel spreadsheet to calculate $a_0$ and its error for each value of $P$. From the graph of $a_0$ against $P$ find the slope and its error and hence obtain a value for $K$ and its error. Compare your answer with the calibrated value on the back of the galvanometer.

[Hint: The expression to calculate $a_0$ can be re-written as $a_0=\frac{a_1^{\frac{3}{2}}}{a_2^{\frac{1}{2}}}$. The error on $a_0$ can therefore be expressed as $\left(\frac{\Delta a_0}{a_0}\right)^2=\frac{9}{4}\left(\frac{\Delta a_1}{a_1}\right)^2+\frac{1}{4}\left(\frac{\Delta a_2}{a_2}\right)^2$. The oscillations can be stopped by shorting the galvanometer by pressing the switch when the oscillation passes the equilibrium point. ]

It can be shown that the charge and current sensitivities are related through the period of the galvanometer:

$$K=2\pi\frac{k}{T} \tag{16.11}$$

Use a stop-watch and measure the period of the galvanometer, and from it the ratio $2\pi/T$. Compare this with your experimentally measured values of $K/k$.

## Procedure

Measurements the internal resistance of the galvanometer and the constant of proportionality $k$.

As shown in Fig. 16.6, turn off the switches $K_1$ and $K_2$, and the voltage across the standard resistance is

$$U_s\approx\frac{R_s}{R_1+R_s}U\approx\frac{R_s}{R_1}U \qquad (R_1\gg R_s) \tag{16.12}$$

At the same time, thecurrent in the galvanometer is usually represented by

$$I_g=\frac{U_s}{R+R_g}\approx\frac{R_sU}{R_1(R+R_g)} \tag{16.13}$$

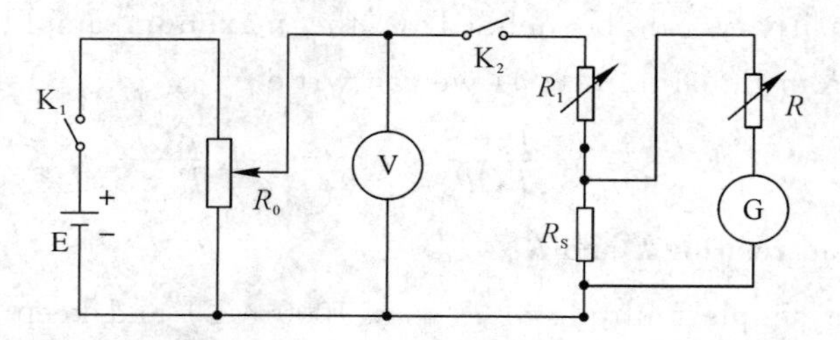


Fig. 16.6 The circuit of the internal resistance of the galvanometer.

Substituting $I_g$ in Formula (16.2), we obtain

$$K_i=\frac{R_sU}{R_1(R+R_g)d} \tag{16.14}$$

1. Assemble electric circuit according to the diagram shown in Fig. 16.6 without connecting the power. All switches must be off.

2. Set resistance R to 500 Ω, adjust resistance $R_0$ to make the voltage $U$ have maximum value (3 V).

3. Change the resistance $R_1$, and make the deflection of the galvanometer around 30 mm. Then, keep the value of $R$ and $U$ (which can be obtained by adjusting $R_0$), and fill Table 16.1 with measured data of $R_i$ and $U_i$.

## Data recording

**Table 16.1 Experiments data.**

External critical resistance: $R_c=$ ____ Ω(theoretical value) $R_c=$ ______ Ω(experimental value)

Standard resistance: $R_s=$ 0.1 Ω; $R_1=$ ________ Ω, $d=$ ____________ mm

| Serial number | 1 | 2 | 3 | 4 | 5 | 6 | 7 | 8 | 9 | 10 |
|---|---|---|---|---|---|---|---|---|---|---|
| $R_i/\Omega$ | 500 | 450 | 400 | 350 | 300 | 250 | 200 | 150 | 100 | 50 |
| $U_i/V$ | | | | | | | | | | |

Revise Equation (16.14) with calculated $R_g$ and $K_i$, and then

$$R=\frac{R_s}{K_i R_1 d}U-R_g \tag{16.15}$$

Use $U$ as horizontal scale and $R$ as vertical scale to draw a relation curve of measured groups of $R$ and $U$. The results can be shown in Fig. 16.7. The intercept in vertical axile $OA$ is the internal resistance $R_g$, and galvanometer constant $K_i$ can be obtained from the expression of curve slope $K=\frac{\Delta R}{\Delta U}=\frac{R_s}{K_i R_1 d}$ in Fig. 16.7.

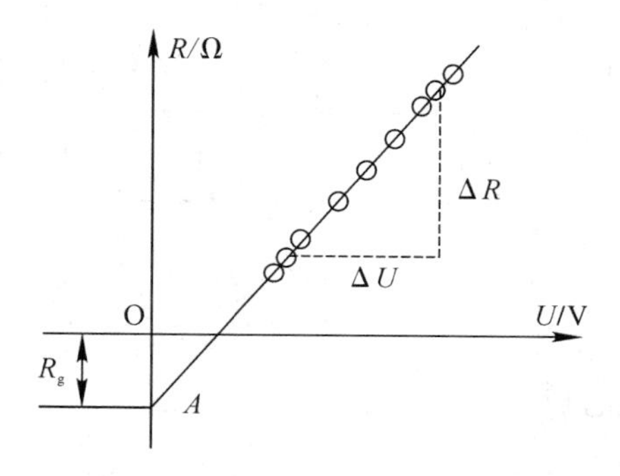


Fig. 16.7 $R\sim U$

# Lab 17 The Magnetic Field of a Solenoid and Ballistic Galvanometer

## Purpose

To measure the magnetic field inside a solenoid and compare the magnetic field with a theoretical value based on the current through the solenoid.

## Apparatus

Ballistic Galvanometer; solenoid; search coil; mutual inductor; DC power supply; standard capacitance box; stopwatch; slide-wire rheostat.

**Ballistic galvanometer**

Ballistic galvanometer is used to measure the migrated electric quantity by a pulsed electric current in a short span of time. The structure of the galvanometer is shown in Fig. 17.1. The ballistic galvanometer has bigger inertia because the deflection coil is flat and wide. In general, the period of sensitive galvanometer is about 1 - 2 seconds, while the period of ballistic galvanometer is about ten seconds.

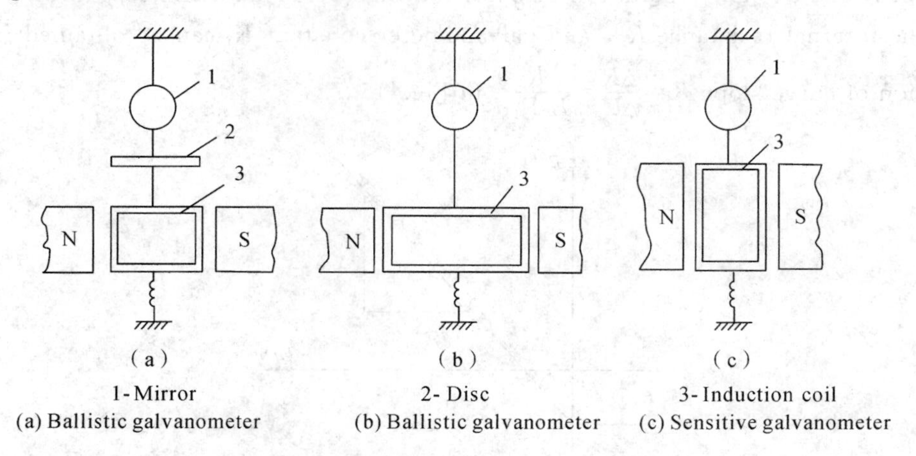


1-Mirror 2-Disc 3-Induction coil
(a) Ballistic galvanometer (b) Ballistic galvanometer (c) Sensitive galvanometer

Fig. 17.1 Structure of galvanometer.

When using ballistic galvanometer to measure electric quantity, the pulse current flowing through the circuit will induce a torque on the galvanometer coil which in turn will deflect the light beam a distance $d$, the magnitude of $d$ being proportional to the current electric quantity:

$$Q = C_b \theta_m = K_b d_m$$

where $\theta_m$ is maximal deflection angle. The constant of proportionality $C_b$ and $K_b$ is defined as the ballistic galvanometer constant which expresses electric quantity through the galvanometer which will produce a torque displacing the beam on the galvanometer by one millimeter.

The value of $K_b$ is not only about the galvanometer but about all-in resistance $R$ in the galvanometer circuit, which can be found by varying $R$ and recording the displacement of the galvanometer beam.

## Theory

When an electric charge is moving, it generates a magnetic field in the space. Since a current is a continuous flow of moving charges, it will generate a magnetic field around it. The magnetic field of a current element is given by Biot-Savart Law:

$$dB = \frac{\mu_0}{4\pi r^3} I dL \times r \tag{17.1}$$

where $I$ is the current; $B$, $L$ and $r$ are, respectively, the magnetic field vector, the length vector of the current element and the displacement vector from the current element to the point of interest. The constant $\mu_0 = 4\pi \times 10^{-7}\,\mathrm{N/A^2}$.

Experimentally, we can measure the magnetic field of a macroscopic current with finite size, instead of an infinitesimal current element. Eq. (17.1) is therefore of theoretically significance only. To obtain the magnetic field of a current of finite size, it is necessary to integrate Eq. (17.1) over the total length of the current, which is somewhat a tedious but straightforward process. Inside a solenoid, the magnetic field along the axis is given by

$$B_X = \frac{\mu N I}{2l} \left\{ \frac{\frac{l}{2} - x}{\left[ \left( \frac{l}{2} - x \right)^2 + r_0^2 \right]^{\frac{1}{2}}} + \frac{\frac{l}{2} + x}{\left[ \left( \frac{l}{2} + x \right)^2 + r_0^2 \right]^{\frac{1}{2}}} \right\} \tag{17.2}$$

where $r_0$ is the radius of the solenoid, $N$ is the number of turns per unit length, $\frac{l}{2} - x$ and $\frac{l}{2} + x$ are the distances from the point of interest, which is the origin of the coordinates in

Fig 17. 2, to the ends of the solenoid. The direction of the magnetic field is along the $X$-axis.

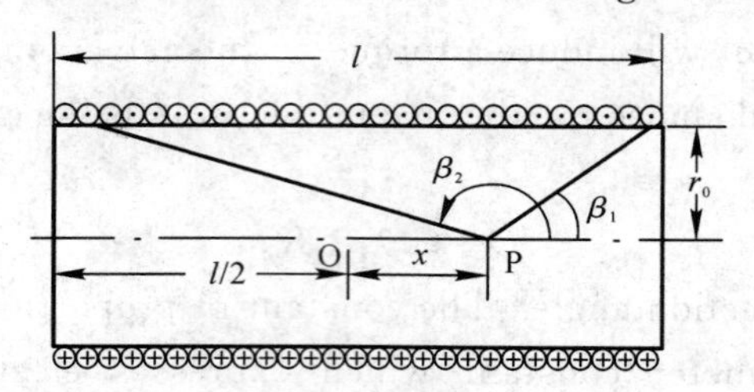


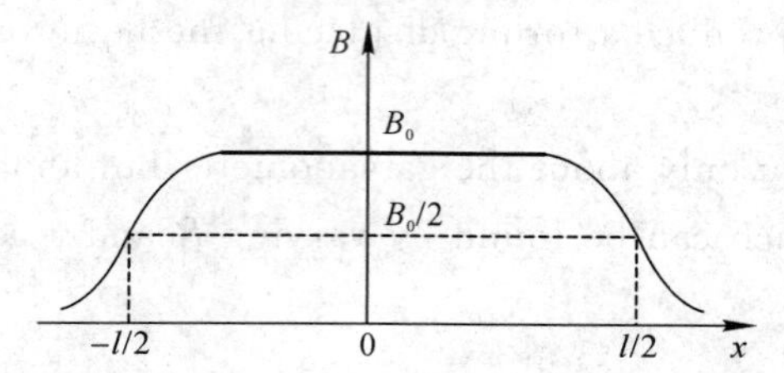


Fig. 17. 2 The magnetic field of a solenoid.

The magnetic field of a solenoid is shown in Fig. 17. 2. Inside the solenoid, the magnetic field is nearly a constant. It starts to decrease dramatically towards the ends of the solenoid. It is sometimes called the "edge effect". The magnetic field quickly reduces to practically zero outside the solenoid. In the central part of the solenoid, if the solenoid is very long, namely, $l \gg r_0$, Eq. (17. 2) can be approximated by a much simpler formula $B=\mu_0 NI$.

In Formula (17. 2), when $x=0$, the magnetic field of a solenoid in the center is

$$B_0=\frac{\mu NI}{(l^2+4r_0^2)^{\frac{1}{2}}} \tag{17. 3}$$

When $x=l/2$ the magnetic field of a solenoid in the ends is

$$B_{l/2}=\frac{\mu NI}{2\ (l^2+4r_0^2)^{1/2}}\approx\frac{\mu NI}{(l^2+4r_0^2)^{\frac{1}{2}}}=\frac{B_0}{2} \quad (l\gg r_0) \tag{17. 4}$$

## Procedure

1. Measure the magnetic field strength

Connect the coil, an ammeter, and the power supply in series, as shown in Fig. 17. 3. E is continuously modulated stabilized DC voltage, A is needle DC galvanometer, $K_1$ and $K_3$ are single-pole single-throw switch (SPST), $K_2$ is single-pole double-throw switch (SPDT), M is mutual inductor, T is the exploring coil within the solenoid S, G is ballistic galvanometer, $R_1$ is resistance box.

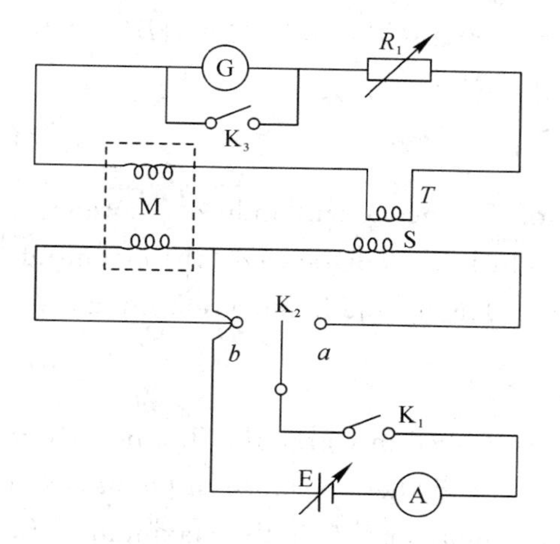


Fig. 17.3 The circuit diagram of measurement.

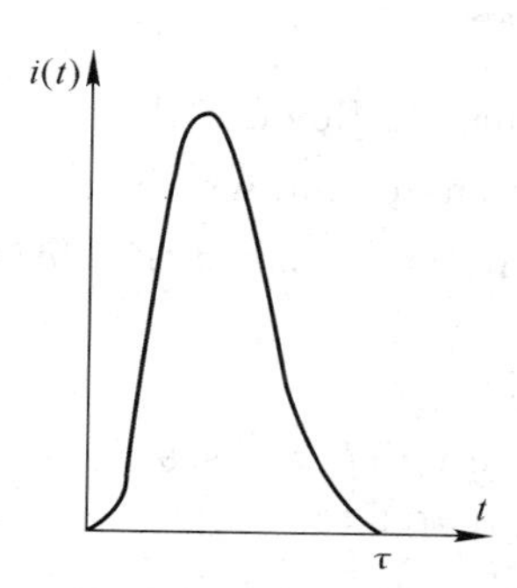


Fig. 17.4 Pulse current.

When the switch $K_2$ is closed to the point "a", the magnetizing current circuit is composed of power supply and solenoid. The subloop is composed of the ballistic galvanometer G, resistance box $R_1$, mutual inductor M, secondary coil and search coil T. The magnetic field in the solenoid changes when electric current flows through it, and then generates the induced electromotive $E(t)$ in the search coil. So the pulse current that changes over time is produced in the measurement circuit which is composed of $R$ and $T$, as in Fig. 17.4. The current satisfies the equation:

$$L\frac{\mathrm{d}i(t)}{\mathrm{d}t}+i(t)R=E(t) \quad \text{or} \quad i(t)=-\frac{L}{R}\frac{\mathrm{d}i(t)}{\mathrm{d}t}+\frac{E(t)}{R} \tag{17.5}$$

where $L$ is the self inductance of the galvanometer loop, $R$ is the all-in resistance of the galvanometer loop which is equal to the sum of galvanometer internal resistance, resistance of the search coil T, secondary resistance of the mutual inductor M and resistance $R_1$. If $n$ is the number of turns in the search coil, and $S$ is its sectional area, the magnetic induction intensity is

$$E(t)=-nS\frac{\mathrm{d}B(t)}{\mathrm{d}t} \tag{17.6}$$

Using Eq. (17.6) in Eq. (17.5):

$$i(t)=-\frac{L}{R}\frac{\mathrm{d}i(t)}{\mathrm{d}t}-\frac{nS}{R}\frac{\mathrm{d}B(t)}{\mathrm{d}t} \tag{17.7}$$

Finally by integrating the pulse current over time of duration, the electric quantity through the coil in the galvanometer loop is $Q=\int_0^{\tau}i(t)\mathrm{d}t=-\frac{L}{R}[i(\tau)-i(0)]-\frac{nS}{R}[B(\tau)-B(0)]$.

We have to first initialize the 0 and $\tau$ points, here $i(\tau)=i(0)=0, B(0)=0, B(\tau)=B(\infty)=B$. So,

$$Q=-\frac{nS}{R}B \tag{17.8}$$

The current flowing through the combination of $R_1$ and $G$ will induce a torque on the galvanometer coil which in turn will deflect the light beam a distance $d$, the magnitude of $d$ being proportional to the current flow: $Q=K_b d_m$. The magnetic induction intensity is

$$B=\frac{RK_b}{nS}d_m \tag{17.9}$$

where Eq. (17.9) uses SI units. The voltage is expressed in volts, the flux in webers, the magnetic field in teslas, the area in square meters, and the resistance in Ohms, $K_b$ is the ballistic galvanometer constant and expressed in C/mm, and $d_m$ is the maximum deflection distance and in mm.

Therefore, if the ballistic galvanometer constant $K_b$ is known we can measure the maximum deflection distance $d_m$, and we can get the magnetic field using Eq. (17.9).

2. Measure the ballistic galvanometer constant $K_b$

When the switch $K_2$ is closed to the point "b", the power supply E and mutual inductor M constitute correction loop. If switch $K_1$ is opened, the current through the primary coil of mutual inductor M will be from $I_0$ to 0 in an instant, In this process, there will be produced a voltage in the secondary coil $E_M=-M\frac{\mathrm{d}i'(t)}{\mathrm{d}t}$ ($M$ is the coefficient of mutual induction). Simultaneously, an impulse current $i(t)=E_M/R$ can be detected in the measuring loop.

With the above theory and deduction, electric quantity through the galvanometer can be expressed as

$$Q=\int_0^{\tau} i(t)\mathrm{d}t=-\frac{M}{R}\int_0^{I_0}\mathrm{d}i'(t)=-\frac{MI_0}{R} \tag{17.10}$$

While switch $K_1$ is turned on, the maximum range for first deflection of the marker of galvanometer is named $d_m'$. When $Q=K_b d_m'$ is substituted into Equation (17.10),

$$K_b=\frac{MI_0}{Rd'_m} \tag{17.11}$$

So

$$RK_b=\frac{MI_0}{d'_m} \tag{17.12}$$

$RK_b$ is expressed in C$\Omega$/mm.

It can be seen from Equation (17. 12) that the ballistic galvanometer constant $K_b$ is related to all-in resistance $R$ of the anometer loop. Therefore, the resistance should be kept the same value when we measure the magnetic induction of solenoid.

3. Experiment content and procedure

(1) Set up the solenoid and connect the circuit as shown in Fig. 17. 3. Ensure switch $K_3$ in closed form.

(2) Connect the power supply of galvanometer lighting and adjust the ray path system. A clear line-of-sight will be found in mirrors under the scale.

(3) Measure $RK_b$(keep the resistance $R_1$ equal to 100 Ω).

① Close switch $K_2$ to "b", turned off switch $K_3$, and turn on switch $K_1$ quickly. Record the maximum deflection distance $d_{m1}$ of the galvanometer cursor and the current $I_0$. Turn off switch $K_1$ quickly. When the galvanometer cursor returns to the equilibrium position, stop it and record the maximum deflection distance $d_{m2}$ of the galvanometer cursor to the other side.

② Adjust the power supply to change the current value $I_0$ through the power circuit, and turn on switch $K_1$ (or turn off) quickly. Record the maximum deflection distance $d_m$ respectively and enter into Table 17. 1. Attention: The value $d_m$ is required to be between 10 and 20 mm.

**Table 17. 1 The data of measurement $RK_b$** $M=$ ________ mH

| $I_0$/mA | $d_{m1}$/cm | $d_{m2}$/cm | $\bar{d}_m$/cm | $k_bR=\frac{MI_0}{\bar{d}_m}$/(CΩ/mm) | $\overline{k_bR}$/(CΩ/mm) |
|---|---|---|---|---|---|
| | | | | | |
| | | | | | |
| | | | | | |

4. Measure the magnetic induction intensity B

① Close switch $K_2$ to "a".

② Set up the search coil which is centered at the axis of the solenoid. The search coil has moved so that its right edge is lined up with the label of solenoid ($x=0$). Turn on switch $K_1$ (or turn off) quickly, and record the maximum deflection distance $d_{\text{left}}$ and $d_{\text{right}}$, respectively.

③ Adjust the search coil position. Measurement of the maximum deflection distance and record are to be filled in Table 17. 2.

**Table 17.2 The data of measurement B.**

$N$=________ turns; $l$=________ m; $r_0$=________ cm

$n$=________ turns; $S$=________ $m^2$; $I$=________ mA

| $x$/cm | 0 | 2 | 4 | 6 | 8 | 10 | 11 | 12 | 13 | 14 | 15 | 16 | 17 | 18 | 19 | 20 |
|---|---|---|---|---|---|---|---|---|---|---|---|---|---|---|---|---|
| $d_{left}$/cm | | | | | | | | | | | | | | | | |
| $d_{right}$/cm | | | | | | | | | | | | | | | | |
| $d_m$/cm | | | | | | | | | | | | | | | | |

## Data Analysis

Plot the experimentally measured $d_m$ along the axis of the solenoid $x$, and calculate the theoretical and experimental values of magnetic field in the middle point $x=0$.

Theoretical value: $B_0=\frac{\mu NI}{(l^2+4r_0^2)^{\frac{1}{2}}}=$________ $T$ ($\mu=4\pi\times10^{-7}\,N/A^2$)

Experiment value: $B=\frac{RK_b}{nS}d_m=$________ $T\left(d_m=\frac{d_{left}+d_{right}}{2}\right)$

Relative error: $E=\left|\frac{B_0-B}{B_0}\right|\times100\%=$________% ($x=0$)

# Lab 18 Linear and Nonlinear Resistors

## Purpose

1. To become familiar with the use of the laboratory DC voltage supply and the use of a digital multimeter (DMM) to measure voltage and current;

2. To learn how to plot I-V characteristics of a linear and nonlinear resistance.

## Apparatus

Digital multimeter (DMM); DC power supply; resistors: 470 Ω, 1000 Ω, 2200 Ω; incandescent lamp; diode.

## Theory

1. Linear Resistors

The current in an electrical circuit depends on the magnitude of the source voltage and the resistance of the various components. For a linear component, the voltage-current ratio is constant. This is usually referred to as Ohm's law, stated as

$$U = I \cdot R \tag{18.1}$$

But for a nonlinear component, the $U/I$ ratio is not constant. So, while the resistance as a particular voltage and current can be obtained from Ohm's law, it will not be constant over the range of currents used. In this experiment you will study the different behaviours of a regular resistor and a light bulb.

2. Non-Linear Resistors

Non-linear conductors such as semiconductors, thermistors and diodes do not have a constant resistance. Instead, we can write the voltage drop as a function of current in the form:

$$U = k \cdot I^n \tag{18.2}$$

where $k$ is a constant.

## Procedure

1. Linear Resistors

(1) Construct the series circuit shown in Fig. 18.1, using $U_s \sim 12$ V dc, $R$ and $R_L$ are both 47 Ω

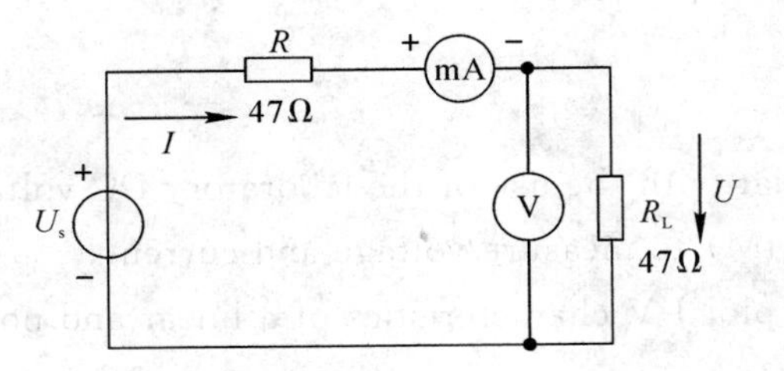


Fig. 18.1 A simple series circuit to study linear resistors.

(2) Starting with $U_s = 0$ V, measure the voltage across $R_L$ and the current through it. By increasing $U_s$ to 0 V, 1 V, 2 V, 3 V, 4 V, 5 V, 6 V, 7 V, 8 V, 9 V, 10 V in suitable steps, obtain a series of voltage and current values.

(3) Plot the measured data points for the incandescent lamp with $I$ on the vertical axis and $U$ on the horizontal axis.

**Table 18.1 The experimental data of linear resistors.**

| $U_s$/V | 0 | 1 | 2 | 3 | 4 | 5 | 6 | 7 | 8 | 9 | 10 |
|---|---|---|---|---|---|---|---|---|---|---|---|
| $I$/mA | | | | | | | | | | | |
| $U$/V | | | | | | | | | | | |
| $R=U/I/\Omega$ | | | | | | | | | | | |

2. Incandescent Lamp

(1) For the circuit in Fig. 18.2, calculate and record $I=U/R$ for 0 V, 2 V, 4 V, 6 V, 8 V, 10 V and 12 V for the incandescent lamp.

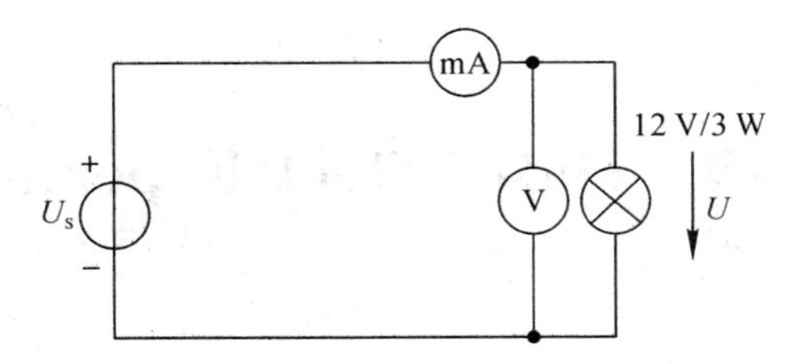


Fig. 18.2 A simple series circuit to study incandescent lamp.

(2) Construct the circuit and adjust the voltage control of the power supply, $U_s$, such that the voltage across $R$ measures each voltage listed at Step (1). Create a table on a separate sheet and record the voltage. Measure and record $I$ for each of these steps in Table 18.2.

(3) Plot the measured data points for the incandescent lamp with $I$ on the vertical axis and $U$ on the horizontal axis. How does this plot compare with the resistor plots?

**Table 18.2 The experimental data of incandescent lamp.**

| $U_s$/V | 0 | 1 | 2 | 3 | 4 | 5 | 6 | 7 | 8 | 9 | 10 | 11 | 12 |
|---|---|---|---|---|---|---|---|---|---|---|---|---|---|
| $I$/mA | | | | | | | | | | | | | |
| $R_{Lamp}/\Omega$ | | | | | | | | | | | | | |
| $U_{Lamp}$/V | | | | | | | | | | | | | |

## Question

Explain why the graph displays this kind of behavior.

# Lab 19 Hall Effect Experiment

The Hall Effect was discovered in 1879 by Edwin Herbert Hall while he was working on his doctoral degree at Johns Hopkins University in Baltimore, Maryland. His measurements of the tiny effect produced in the apparatus he used were an experimental tour de force, accomplished 18 years before the electron was discovered.

## Purpose

After successfully completing this laboratory workshop, including the assigned reading, the lab problems, and any required reports, the students will be able:

1. To determine the semiconductor type from the polarity of the Hall voltage, knowing the orientation of all fields and currents in the experimental arrangement;

2. To calculate the carrier concentration and mobility from the magnitude of the Hall voltage and the known experimental variables (magnetic field and sample resistance);

3. To explain the response of the charge carriers in a material to a magnetic field;

4. To predict the Hall voltage which would develop under a given set of Hall experiment variables.

## Apparatus

Hall Effect measurement apparatus; Hall Effect measurement monitor; wires.

## Theory

1. Hall Effect

The Hall Effect is the production of a voltage difference (the Hall voltage) across an electrical conductor, transverse to an electric current in the conductor and a magnetic field perpendicular to the current. It can be seen in Fig. 19.1.

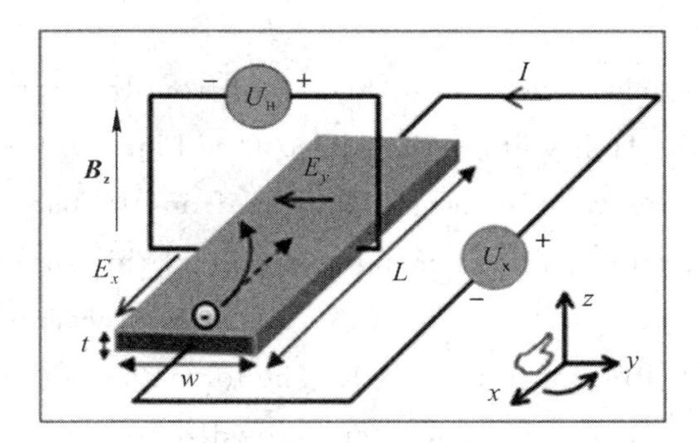


Fig. 19.1 The schematic diagram of the Hall effect.

The Hall Effect occurs because a charged particle moving in a magnetic field is subject to the Lorentz force given by

$$F_{\text{Lorentz}} = q\boldsymbol{v} \times \boldsymbol{B} \tag{19.1}$$

where $q$ is a signed quantity representing carrier charge, $\boldsymbol{v}$ is the particle velocity vector, and $\boldsymbol{B}$ is the vector magnetic field. The basic Hall measurement is performed on a semiconductor bar with an electric field applied along its long axis, and magnetic field applied perpendicular to it.

2. Hall Voltage

Examine the sample shown in Fig. 19.1. A voltage $V$ is applied, giving rise to a field $E_x = V/L$. If the sample is n-type, the majority of carrier electrons will move opposite the applied electric field, right to left ($-x$). The $\boldsymbol{v} \times \boldsymbol{B}$ cross product is in the positive $y$-direction for a $\boldsymbol{B}$-field directed upward on the page. The carrier charge in this case is negative so the force is actually in the $y$-direction. This force causes the majority of carrier electrons to be pushed towards the front edge of the sample. The entire sample remains neutrally charged as the positive charges of the donor ions are now uncompensated at the back. There is, however, a gradient of charge increasing from back to front giving rise to a second electric field perpendicular both to the externally applied electric field $E_x$ and the $\boldsymbol{B}$-field. This new electric field opposes further accumulation of electrons (it could be viewed as rejection by electrons already there). The system reaches equilibrium.

When the force applied on carriers by the second electric field $E_y$ equals and opposes the force due to the $\boldsymbol{B}$-field,

$$qE_y = q\,|v_x \cdot B_z| \tag{19.2}$$

$E_y$ gives rise to a voltage which can be measured from the front to the back face of the sample. This is called the Hall voltage $V_H = E_y w$ (see Fig. 19.1 for the definition of $w$). In this n-type sample, the voltage measured from front to back in the sample will be negative. If any of the sign definitions change, however, this sign may change too.

Now consider a p-type sample. Here the majority of carrier holes move with the applied electric field, left to right in Fig. 19.1. The force due to the magnetic field $q\boldsymbol{v} \times \boldsymbol{B}$ is in the $y$-direction and once more carriers are crowded to the front face of the sample resulting in an electric field.

This time, however, because the carriers are positive, the Hall voltage measured from front to back on the sample will be positive. Thus, the majority carrier type determines the sign of the Hall voltage. The velocity we have been discussing is the carrier drift velocity and is related to the current by

$$I_x = qnv_xA \qquad \text{or} \qquad I_x = qpv_xA \tag{19.3}$$

For n-type or p-type sample, remember that $q$ is a signed quantity. In Equation (19.3) $I_x$ is the current in the $x$-direction due to the applied electric field, $n$ or $p$ is the carrier concentration per $cm^3$, and $A$ is the cross sectional area of the sample in $cm^2$ (width times thickness: $wt$). The quantity $v$ can easily be solved for and the result is substituted in Equation (19.2), resulting in

$$E_y = \frac{-I_x B_z}{|q|\,nA}\ \text{(n-type)}, \qquad E_y = \frac{+I_x B_z}{|q|\,pA}\ \text{(p-type)} \tag{19.4}$$

in which the sign of the charge carrier is now expressed explicitly.

If we substitute the product $wt$ for $A$ in Equation (19.4), where $w$ is the sample width and $t$ the sample thickness, we can multiply Equation (19.4) by the width to get the Hall voltage:

$$U_H = wE_y = \frac{-I_x B_z}{|q|\,nt}\ \text{(n-type)}, \qquad U_H = wE_y = \frac{+I_x B_z}{|q|\,pt}\ \text{(p-type)} \tag{19.5}$$

Since $U_H$, $B_z$, $I_x$, $t$ and $q$ are all known (by measurement), it is possible to solve for the carrier concentration $n$ or $p$, and determine whether the sample is n-type or p-type.

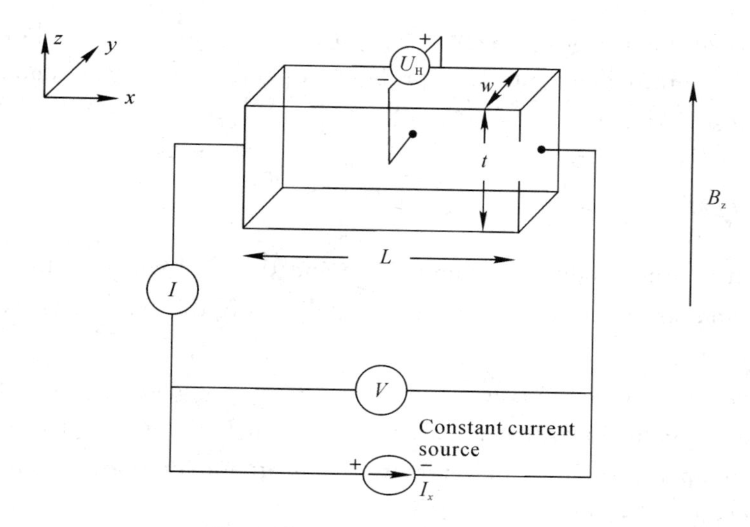


Fig. 19.2 Hall voltage circuit.

3. Hall Coefficient

The Hall coefficient is defined as the ratio of the induced electric field to the product of the current density and the applied magnetic field. It is a characteristic of the material from which the conductor is made, since its value depends on the type, number, and properties of the charge carriers that constitute the current. A useful measurement concept is Hall coefficient, which is defined as

$$R_H = \frac{E_y}{J_x B_z} \tag{19.6}$$

Writing $R_H$ in terms of measurable quantities, we get

$$R_H = \frac{U_H t}{I_x B_z} \tag{19.7}$$

If we substitute the value of $U_H$ from Equation (19.5) into Equation (19.7), we can see

$$R_H = \frac{-1}{|q|n} \quad \text{(n-type)}, \qquad R_H = \frac{+1}{|q|p} \quad \text{(p-type)} \tag{19.8}$$

The above analysis relies upon the idea that all carriers travel with the drift velocity.

4. Hall Sesitivity

The Hall Effect can be used to measure magnetic fields with a Hall probe. If we substitute Equation (19.3) into Equation (19.5), we can see

$$U_H = K_H I B \tag{19.9}$$

where $K_H$ is called Hall sesitivity, which is the maginitude of the Hall voltage field intensity and one operating current. Hall probes can be made very small and are convenient and accurate to use. And then,

$$B=\frac{U_H}{K_H I} \tag{19.10}$$

5. Hall Mobility

Hall mobility is the product of conductivity and the Hall constant for a conductor or semiconductor and is a measure of the mobility of the electrons or holes in a semiconductor.

Thermo magnetic effects: Electrical and thermal phenomena occurring when a conductor or semi-conductor which is carrying a thermal current is placed in a magnetic field. Let the temperature be transverse to the magnetic field $\boldsymbol{B}$. Then the following transverse effects are observed:

(1) Ettingshausen effect

When a metal strip is placed with its plane perpendicular to a magnetic field $\boldsymbol{B}$ and an electric current $I$ is sent longitudinally through the strip, cornensponding points on opposite edges of the strip have different temperatures, whichs causes the potential difference $U_E$ to change with the directions of $B$ and $I$. This phenomenon is called Ettingshausen effect.

(2) Nernst effect

When a conductor is placed in a magnetic field $\boldsymbol{B}$ and an electric current $I$ flows through the conductor perpendicular to the field, a temperature gradient arises in the direction of the current. There is current deflection in the condition of the magnetic field so that there produces additional potential difference $U_N$, which is related only with the direction of $\boldsymbol{B}$.

(3) Right-Leduce effect

A temperature gradient exists along $E_H$ direction. If a magnetic field is applied at right angles to the direction of a temperature gradient in a conductor, a new temperature gradient is produced perpendicular to both the direction of the original temperature gradient and to the magnetic field. The new temperature gradient induces another additional potential difference $U_R$, which is only related with the direction of $\boldsymbol{B}$.

(4) A potential difference $U_o$ change along the direction of the electron movement. When a current flows through a Hall electrode, there exists potential difference $U_o$ because the two Hall electrodes can hardly reach the same potential. The potential difference is only with the direction of the current $I$.

In order to reduce the influence of the above additional effects, the common method is to switch the directions of the magnetic field $\boldsymbol{B}$ and the current $I$. The sign of positive and negative denote the reverse directions.

$$\begin{array}{lll} +B & +I & U_1=+U_H+U_E+U_N+U_R+U_o \\ +B & -I & U_2=-U_H-U_E+U_N+U_R-U_o \\ -B & -I & U_3=+U_H+U_E-U_N-U_R-U_o \\ -B & +I & U_4=-U_H-U_E-U_N-U_R+U_o \end{array} \tag{19.11}$$

Then

$$\frac{1}{4}(U_1-U_2+U_3-U_4)=U_H+U_E\approx U_H \tag{19.12}$$

So, there remains only $U_H$ and $U_E$ in the last expression. $U_E$, which is a tiny quantity, can be neglected.

## Procedure

1. Observe the measurement apparatus and the monitor, and check its circuit.

(1) Hall effect measurement apparatus, which includes solenoid, two-dimensional motion rule, three double-throw switches, Hall strip and its leads, is shown in Fig. 19.3. A solenoid is a three-dimensional coil. In physics, the term solenoid refers to a loop of wire, often wrapped around a metallic core, which produces a magnetic field when an electric current is passed through it. One Hall strip is fixed on one probe controlled by horizontal and vertical knobs. A Hall strip has two pairs of electrodes (as shown in Fig. 19.4), which connect with the two switches for the operation current $I_s$ and the Hall potential $U_H$, respectively.

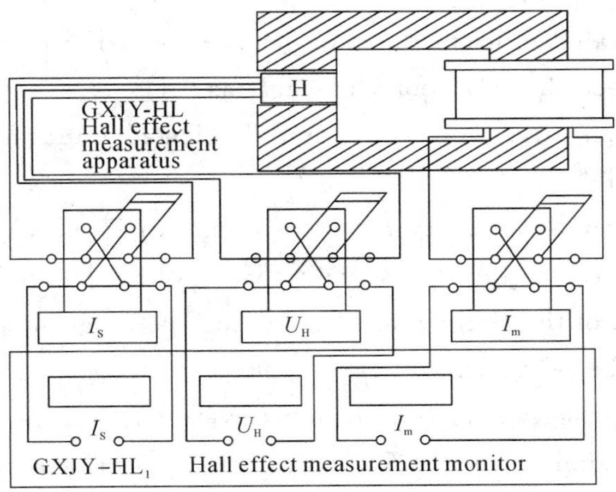


Fig. 19.3  Sketch map of Hall effect measurement apparatus.

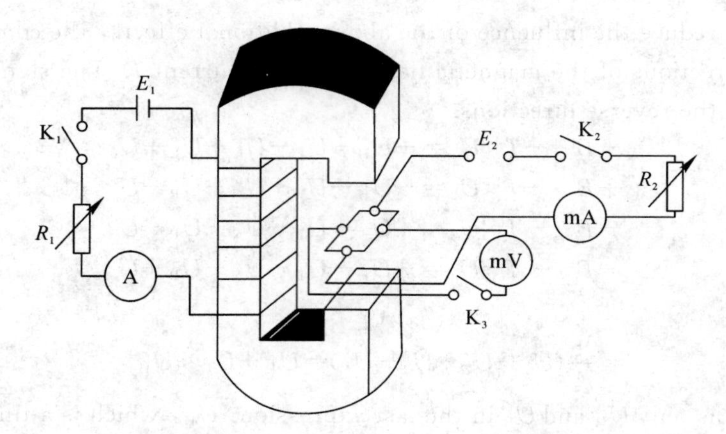


Fig. 19.4 Arrangement diagram of Hall effect experiment.

The parameters, such as $K_H$, the size (b, d, L) of the Hall strip, are generally offered by laboratory assistants. The direction of excitation current $I_m$ (to produce the magnetic field of solenoid), operation current $I_s$ and Hall potential $U_H$ are changed by the three double-pole double-throw switches, respectively.

(2) Hall effect measurement monitor includes the display windows for the excitation current $I_m$, the operation current $I_s$ and the Hall potential $U_H$. The measurement range of the excitation current $I_m$ is 0 - 1000 mA. The measurement range of the operation current $I_s$ is 0 - 100 mA. The display windows of the two currents are located on the right. The measurement range of the Hall potential $U_H$ is $-199.9 - 199.9$ mV. The Hall potential is displayed on the left window.

2. Measure the Hall coefficient $R_H$ and calculate the carrier density $n$.

(1) Adjust and keep the excitation current $I_m$ as 0.50 A.

(2) Move the Hall strip and center in the solenoid interspaces through the two-dimensional motion rule.

(3) Adjust the operation current $I_s$ from 1.0 mA to 10.0 mA with an increment of 1.0 mA. Measure and record the corresponding Hall voltage $U_H$ in Table 19.1.

(4) Plot the curve of the Hall voltage $U_H$ and the operation current $I_s$. Solve the slope $k=U_H/I_s$. Conclude the relationship.

(5) Calculate the magnet field $B$ by $B=U_H/(K_H \cdot I_s)$.

(6) Calculate the Hall coefficient $R_H$ and the carrier density $n$.

(7) Judge the type of the semiconductor, n-type or p-type.

3. Study the relationship between the Hall voltage $U_H$ and the excitation current $I_m$.

(1) Adjust and keep operation current $I_s$ as 10.0 mA.

(2) Move the Hall strip and center in the solenoid interspaces through the two-dimensional motion rule.

(3) Adjust the excitation current $I_m$ from 0.1 A to 1.0 A with an increment of 0.1 A. Measure and record the corresponding Hall voltage in Table 19.2.

(4) Plot the curve of the Hall voltage $U_H$ and the excitation current $I_m$. Conclude the relationship.

4. Plot the relationship curve of the magnetic field $B$ and the excitation current $I_m$ and $B \sim I_s$, and conclude the relationship.

**Table 19.1 Measurement of $U_H - I_s$** ($I_m = 500$ mA).

| $I_s$/mA | 1.0 | 2.0 | 3.0 | 4.0 | 5.0 | 6.0 | 7.0 | 8.0 | 9.0 | 10.0 |
|---|---|---|---|---|---|---|---|---|---|---|
| $(+B, +I)U_1$/mV | | | | | | | | | | |
| $(+B, -I)U_2$/mV | | | | | | | | | | |
| $U_H$/mV | | | | | | | | | | |
| $B/T$ | | | | | | | | | | |

**Table 19.2 Measurement of $U_H - I_m$** ($I_s = 10$ mA)

| $I_m$/A | 0.10 | 0.20 | 0.30 | 0.40 | 0.50 | 0.60 | 0.70 | 0.80 | 0.90 | 1.00 |
|---|---|---|---|---|---|---|---|---|---|---|
| $(+B, +I)U_1$/mV | | | | | | | | | | |
| $(+B, -I)U_2$/mV | | | | | | | | | | |
| $U_H$/mV | | | | | | | | | | |
| B/T | | | | | | | | | | |

## Questions

(1) What is the drift velocity of the carriers?

(2) Please induce the Hall coefficient $R_H$ unit.

# Lab 20 Mapping the Electric Field

## Purpose

To map the electric and potential fields resulting from three different configurations of charged electrodes—rectangular, concentric, and circular.

## Apparatus

Electrostatic field plotter.

## Theory

An *electric field* is a region in which forces of electrical origin are exerted on any electric charges that may be present. If a force $\boldsymbol{F}$ acts on a charge $q$ at some particular point in the field, the electric field strength $\boldsymbol{E}$ at that point is defined as the force per unit charge, and the magnetite is given by

$$\boldsymbol{E}=\frac{\boldsymbol{F}}{q} \tag{20.1}$$

$\boldsymbol{E}$ is a vector quantity: it has both magnitude and direction and has the unit of newtons per coulomb. The field is set up by electric charges somewhere in the surrounding space. For continuous charge distributions, it is much easier to analyze electric forces using the field concept than using Coulomb's Law for the forces between point charges.

If we measure the field (magnitude and direction) at enough points around the charge distribution, we could make a map of the electric field lines. These lines show the direction of the electric force at each point.

However, instead of measuring the electric field directly, we will measure and map the *electric potential* around the charge distribution.

### Electric Potential

To place a positive test charge in an electric field we must do work against the field, since the field tries to push the charge away. As the electric force is conservative, the work we do in placing the charge in the field is stored as potential energy.

*Electric potential* is defined as the work done per unit charge to move the charge into an electric field. Like potential energy, electric potential is measured relative to some reference position, so that we define it as $\Delta V = W/q$, where $\Delta V$ is the change in electric potential in going from the reference point to the point in question, $W$ is the work done (or change in potential energy) and $q$ is the charge. The unit of electric potential is the volt, which is equivalent to a joule per coulomb. We usually use the Earth itself as our reference point, and take the electric potential of the Earth to be zero.

**Equipotential Lines**

A *line of force* is defined as the path traversed by a free test charge as it moves from one point to another in the field. Fig. 20.1 shows several possible paths that a test charge might take in going from the positively charged body A to the negatively charged body B. The relative magnitude of the field intensity is indicated by the spacing of the lines of force and the arrows indicate the direction.

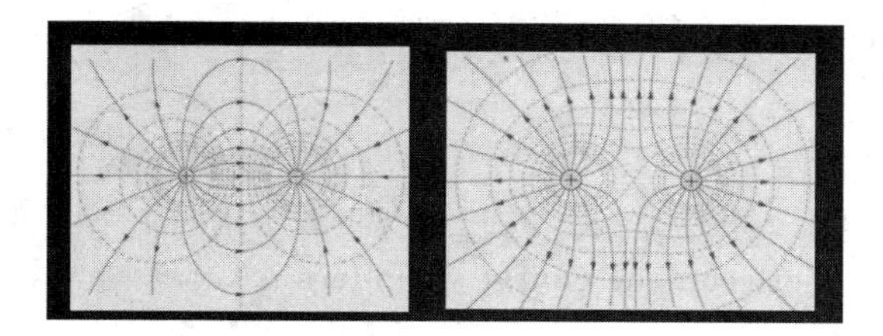

Fig. 20.1 The field around two cylindrical electrodes.

Thus a point in an electric field or in an electric circuit can be characterized by the electric potential, or simply the potential, there. A quantity of charge placed at that point has potential energy equal to the voltage there times the amount of charge.

We can measure electric potential directly using a voltmeter, then map the potential field by connecting points that are at the same electric potential. Moving a charge along an equipotential line requires no work, since the energy of the charge does not change.

To find the direction of the electric field, we make use of the fact that the equipotential lines must always be perpendicular to the electric field lines. This is because the electric field lines show the direction of maximum decrease in the potential.

One can readily see that many equipotential lines, or surfaces, are possible in an electric field. Experimentally, it is much easier to trace the path of equipotential lines than to trace the lines of force. When a network of the equipotenial lines have been mapped out, the lines of force, being everywhere normal in the equipotential lines, can be readily plotted.

Any electric field due to an electrostatic field or a stable current field can be determined from potential differences, with the vector equation:

$$E_l = -\frac{dU}{dl} \tag{20.2}$$

where $E_l$ is the component of the electric field in the direction of the infinitesimal displacement $dl$. The quantity of the vector $\boldsymbol{E}$ is determined by the gradient of the potential (grad $U$) which changes most rapidly.

(1) Coaxial cylindrical cable and its electric field

An example is the electric field due to two opposite charged coaxial cylinders with radii $r_a$(inner) and $r_b$(outer) written by Gauss's law as

$$E_r = \frac{\lambda}{2\pi\varepsilon_0} \cdot \frac{1}{r} \tag{20.3}$$

where $\lambda$ is the length charge density on the cylindrical electrodes, $r$ is the radius of a cylindrical equipotential surface between the electrodes, and $r_a < r < r_b$. The potential on the surface is

$$U_r = \int_r^b E_r \mathrm{d}r = \frac{\lambda}{2\pi\varepsilon_0} \cdot \ln\frac{b}{r} \tag{20.4}$$

The potential difference between the charge cylinders is

$$U_{r_a r_b} = \int_{r_a}^{r_b} E_r \mathrm{d}r = \frac{\lambda}{2\pi\varepsilon_0} \cdot \ln\frac{r_b}{r_a} \tag{20.5}$$

(2) Coaxial cylindrical cable and the electric currents

If there is homogenous conductor with conductivity filled between the two cylindrical electrodes, the electric current density at the direction of radius is

$$j_r = \sigma E_r = \frac{\sigma\lambda}{2\pi\varepsilon_0} \cdot \frac{1}{r} = \frac{U_{ab}}{\ln\frac{r_b}{r_a}} \cdot \frac{\sigma}{r} \tag{20.6}$$

And the current on the cylinders of length $d$ is

$$I = d\int_L j_r \mathrm{d}l = j_r d\, 2\pi r = 2\pi\frac{U_{ab}\sigma}{\ln\frac{r_b}{r_a}} \cdot d \tag{20.7}$$

Experimentally, it is also easy to obtain the conductivity from the current and the voltage charged on the electrodes:

$$\sigma = \frac{IU_{ab}d}{2\pi} \cdot \ln\frac{r_b}{r_a} \tag{20.8}$$

Table 20. 1 lists some electric fields due to electrodes.

**Table 20. 1 Electrodes and electric fields.**

| | Electrode | Electric field |
|---|---|---|
| Coaxial cylindrical cable | 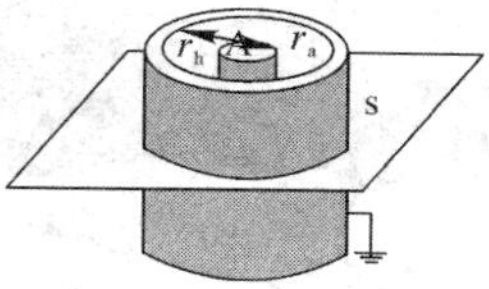 $r_b$ A $r_a$ S | 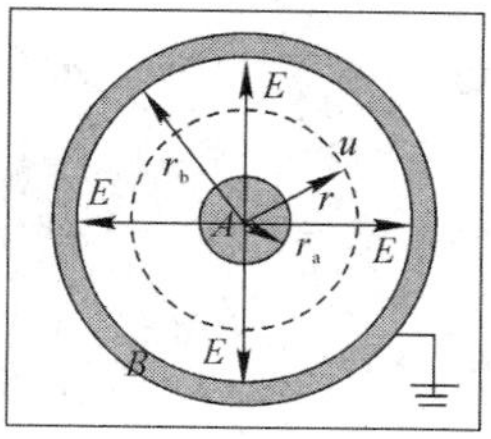 E $r_b$ u E r A $r_a$ E B E |
| Parallel wires | 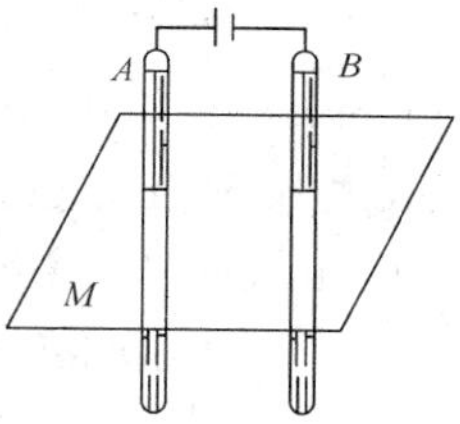 A B M | 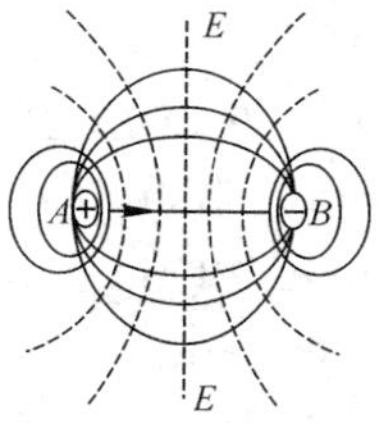 E A B E |
| Parallel plates | 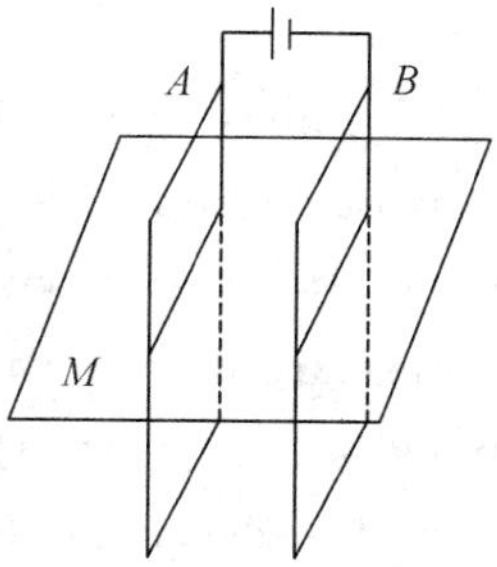 A B M | 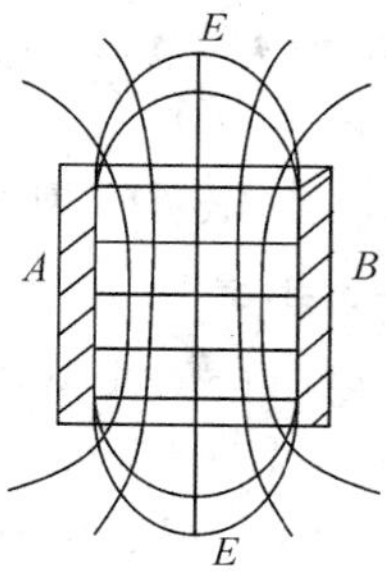 E A B E |

## Procedure

As shown in Fig. 20. 2, GVZ-4 Electrostatic field plotter is a dedicated device to measure and display electric field with electric conductive crystallite. It in cludes a manipulative power and a measuring box. On the box, there are four plate electrodes inside and one probe as necessary. Measurement of concentric circles electrode is shown by polar coordinates and results of other kinds of electrode are shown by rectangular coordinates on the screen.

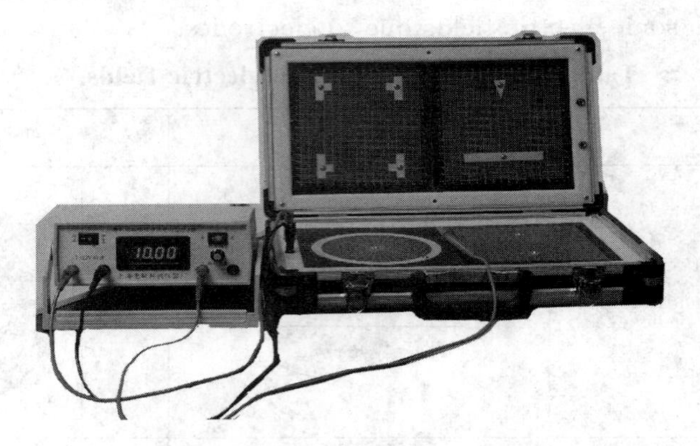


Fig. 20. 2 GVZ-4 Electrostatic field plotter.

The red wire from the power with a dedicated stabilized voltage supply and an electrostatic field need to be connected to the red point with a sign of "+" on the left side of the plate of the box. Similarly, the black wire from the power needs to be connected to the black point with a sign of " − " on the left side of the plate of the box. The measurement can be practiced after the other side of the red test probe is plugged in the jack on the lower right of the power panel.

To conduct an experiment, the lead lines of electrodes must be connected with the binding posts after 10 V direct current main from the power is provided. When the pilot lamp works after the power is turned on, the check switch on top left of the power panel is turned to the position of revise. If a measure point on the crystallite plate is found by the test probe, related location of the point should be marked on recording paper. Then, an equipotential line can be painted if several measure points with parallel electric potential are found when the test probe is moved to different places of the crystallite plate.

1. Describing static electric field distribution of the coaxial line

As shown in Fig. 20. 3, connect the electrodes of the crystallite plate with the DC stabilized power and move the test probe to map equipotential lines of the coaxial line with 1 - 7 V voltage. That potential difference between every two adjacent equipotential lines is 1 V is needed. And, concentric circles with different radii, named $r$, from those points of each equipotential line to the original point can be obtained. The whole distribution diagram of electric field is finished after electric field lines are drawn with cross principle between its equipotential lines and related directions of electric field intensity are shown clearly. Then, the curve related relative electric potential $\frac{U'_r}{U_a}$ and $\ln r$ need to be drawn on coordinate paper and compared with theoretical results.

2. Describing stationary potential distribution of electrode model of parallel lines

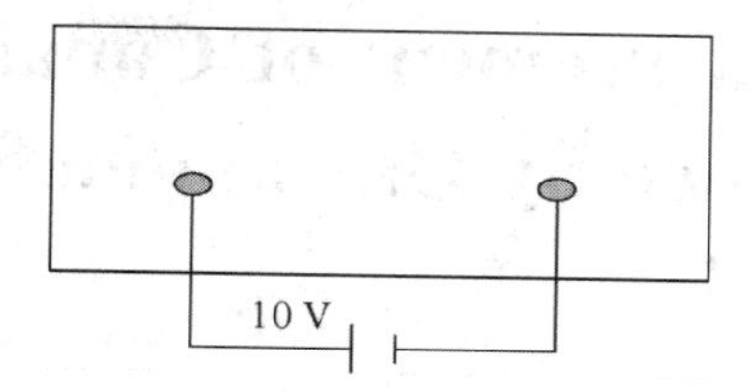


Fig. 20. 3 Electrode model of parallel lines.

As shown in Fig. 20. 3, when each equipotential point on the plate is found by the test probe, a related sign needs to be marked on the recording paper. A series of equipotential points, always more than 10 points of each line, are detected of each voltage interval of 1 V. Six equipotential lines can be obtained when the voltage changes from 3 V to 8 V. To measure more points around each electrode are recommended.

## Questions

1. If the space surrounding the electrode configuration were completely nonconducting, explain how your observations with probes would be affected.

2. If the supply voltage $U_a$ is doubled do shapes of equipotential lines and power lines change, and do distribution of electric field intensity and electric potential change? Why?

3. Is there linear relation between $U'_r$ and $\ln r$ in the model of coaxial line? Why?

# Lab 21 The Measurement of Capacitance and High Resistance by Galvanometer

## Purpose

1. To determine the capacitance by the comparative method using a Ballistic galvanometer;

2. To determine the high resistance by method of leakage of charge using a Ballistic galvanometer.

## Apparatus

Ballistic galvanometer; solenoid; search coil; mutual inductor; DC power supply; the standard capacitance box; stopwatch; slide-wire rheostat.

**Introduction to ballistic galvanometer**

Ballistic galvanometer is used to measure the migrated electric quantity by a pulsed electric current in a short span of time. The structure of the galvanometer is shown in Fig. 21.1. The ballistic galvanometer has bigger inertia because the deflection coil is flat and wide. In general, the period of sensitive galvanometer is about 1 - 2 seconds, while the period of ballistic galvanometer is about ten seconds.

When using ballistic galvanometer to measure electric quantity, the pulse current flowing through the circuit will induce a torque on the galvanometer coil which in turn will deflect the light beam a distance $d$, the magnitude of $d$ being proportional to the current electric quantity:

$$Q=C_b\theta_m=K_b d_m$$

where $\theta_m$ is maximal deflection angle. The constant of proportionality $C_b$ and $K_b$ is defined as the ballistic galvanometer constant which expresses electric quantity through the galvanometer which will produce a torque displacing the beam on the galvanometer by one millimeter.

The value of $K_b$ is not only about the galvanometer but about all-in resistance $R$ in the galvanometer circuit, which can be found by varying $R$ and recording the displacement of the galvanometer beam.

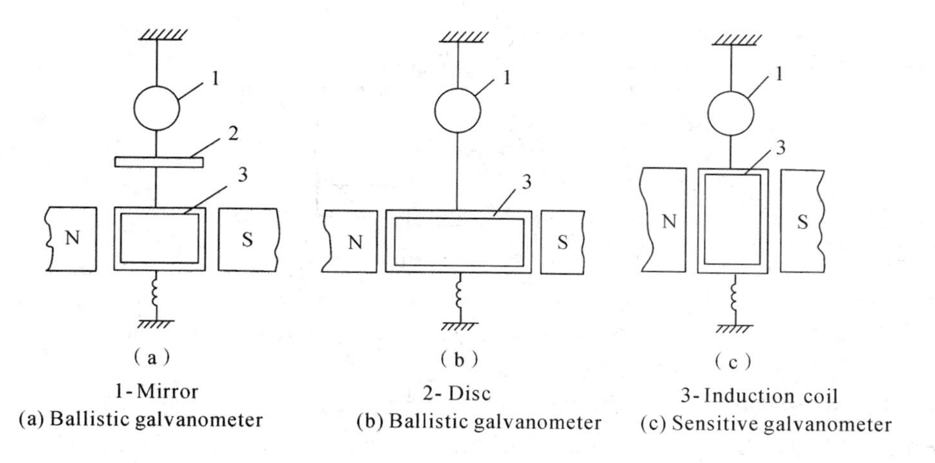


Fig. 21.1 Structure of galvanometer.

## Theory

1. The measurement of capacitance

Measuring a capacitor's capacity with a ballistic galvanometer is as shown in Fig. 21.2. The standard capacitor is charged by power supply when Switch $K_2$ is closed to "a" and $K_3$ closed to "$C_0$". The charge capacity of the capacitor is

$$Q_0 = C_0 U \tag{21.1}$$

When Switch $K_2$ is closed to "b", the charge capacity of the standard capacitor is equal to $Q_0$. Then Switch $K_3$ is closed to "$C_x$" and repeat the steps above. The voltage being kept constant, there is

$$Q_x = C_x U \tag{21.2}$$

So by direct substitution into Equation (21.1):

$$C_x = \frac{Q_x}{Q_0} C_0 \tag{21.3}$$

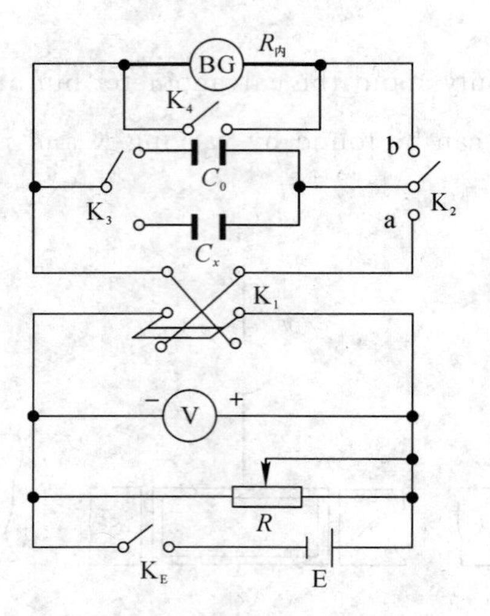


Fig. 21.2 The circuit diagram of measuring a capacitor's capacitance.

2. The measurement of high resistance

In Fig. 21.3, set the capacitance of the capacitor to $C_0$ and charge it to $U_0$. The electricity of the capacitance $Q_0 = C_0U_0$, and the rule of charge and discharge can be expressed as

$$Q = Q_0 \exp\left(-\frac{t}{RC_0}\right) \tag{21.4}$$

where $C_0$ and $Q_0$ are known quantities. So we can get the experimental resistance if the electric charges stored in the capacitor can be measured. Take natural log of Eq. (21.4):

$$\ln Q_0 - \ln Q = \frac{t}{RC_0} \tag{21.5}$$

Substituting Eqs. $Q_0 = C_0U_0$ and $Q = k_b d$ into Eq. (21.5), we obtain

$$\ln d = \ln \frac{C_0U_0}{k_b} - \frac{t}{RC_0} \tag{21.6}$$

Eq. (21.6) shows $\ln d$ depends linearly on $t$, the resistance ($R$) can be obtained with the slope of line $\ln d \sim t$, which is based on the values of the deflection $d$ and the value of discharge time $t$.

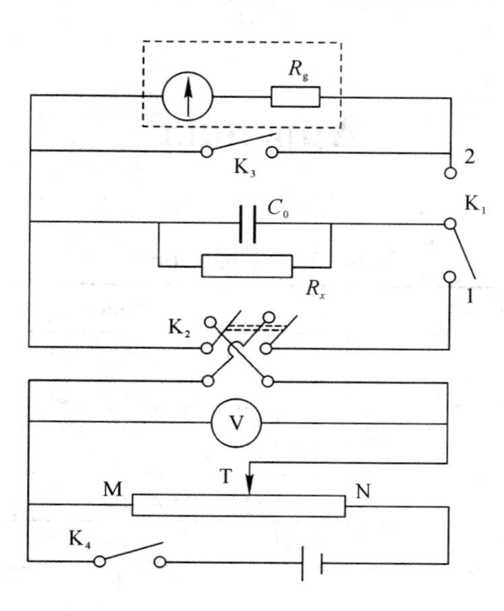


Fig. 21.3 The circuit diagram of measuring the high resistance.

## Procedure

1. Connect the voltmeter, the capacitor, galvanometer and the power supply in series, as shown in Fig. 21.2.

2. Turn on Switch $K_2$ to "a" recharging the capacitor and then to "b" and the capacitor discharges the galvanometer. Measure $d_x$.

3. Switch $K_3$ is closed to $C_0$; keep the value of $R$ and $U$ unchanged, adjust the standard capacitance box to get the right $C_0$ and make $d_x/d_0=1$.

4. Calculate the capacitance under test using the formula $C_x=C_0 \cdot d_x/d_0$.

5. Set up the circuit shown in Fig. 21.3, make a series of measurements of the deflection $d$ as $t$ is increased. Plot line $\ln d \sim t$ and calculate the high resistance.

# Appendix

**Table 1 SI base units**

| Base quantity | Name | Symbol |
|---|---|---|
| length | meter | m |
| mass | kilogram | kg |
| time | second | s |
| electric current | ampere | A |
| thermodynamic temperature | kelvin | K |
| amount of substance | mole | mol |
| luminous intensity | candela | cd |

**Table 2 Examples of SI derived units**

| Derived quantity | Name | Symbol |
|---|---|---|
| area | square meter | $m^2$ |
| volume | cubic meter | $m^3$ |
| speed, velocity | meter per second | m/s |
| acceleration | meter per second squared | $m/s^2$ |
| wave number | reciprocal meter | $m^{-1}$ |
| mass density | kilogram per cubic meter | $kg/m^3$ |
| specific volume | cubic meter per kilogram | $m^3/kg$ |
| current density | ampere per square meter | $A/m^2$ |
| magnetic field strength | ampere per meter | A/m |
| amount-of-substance concentration | mole per cubic meter | $mol/m^3$ |
| luminance | candela per square meter | $cd/m^2$ |
| mass fraction | kilogram per kilogram, which may be represented by the number 1 | kg/kg=1 |

**Table 3 SI derived units with special names and symbols**

| Derived quantity | Name | Symbol | Expression in terms of other SI units | Expression in terms of SI base units |
|---|---|---|---|---|
| plane angle | radian[a] | rad | — | $m \cdot m^{-1}=1$ [b] |
| solid angle | steradian[a] | sr [c] | — | $m^2 \cdot m^{-2}=1$ [b] |
| frequency | hertz | Hz | — | $s^{-1}$ |
| force | newton | N | — | $m \cdot kg \cdot s^{-2}$ |
| pressure, stress | pascal | Pa | $N/m^2$ | $m^{-1} \cdot kg \cdot s^{-2}$ |
| energy, work, quantity of heat | joule | J | $N \cdot m$ | $m^2 \cdot kg \cdot s^{-2}$ |
| power, radiant flux | watt | W | J/s | $m^2 \cdot kg \cdot s^{-3}$ |
| electric charge, quantity of electricity | coulomb | C | — | $s \cdot A$ |
| electric potential difference, electromotive force | volt | V | W/A | $m^2 \cdot kg \cdot s^{-3} \cdot A^{-1}$ |
| capacitance | farad | F | C/V | $m^{-2} \cdot kg^{-1} \cdot s^4 \cdot A^2$ |
| electric resistance | ohm | Ω | V/A | $m^2 \cdot kg \cdot s^{-3} \cdot A^{-2}$ |
| electric conductance | siemens | S | A/V | $m^{-2} \cdot kg^{-1} \cdot s^3 \cdot A^2$ |
| magnetic flux | weber | Wb | $V \cdot s$ | $m^2 \cdot kg \cdot s^{-2} \cdot A^{-1}$ |
| magnetic flux density | tesla | T | $Wb/m^2$ | $kg \cdot s^{-2} \cdot A^{-1}$ |
| inductance | henry | H | Wb/A | $m^2 \cdot kg \cdot s^{-2} \cdot A^{-2}$ |
| Celsius temperature | degree Celsius | °C | — | K |
| luminous flux | lumen | lm | $cd \cdot sr$[c] | $m^2 \cdot m^{-2} \cdot cd=cd$ |
| illuminance | lux | lx | $lm/m^2$ | $m^2 \cdot m^{-4} \cdot cd=m^{-2} \cdot cd$ |
| activity (of a radionuclide) | becquerel | B q | — | $s^{-1}$ |
| absorbed dose, specific energy (imparted), kerma | gray | Gy | J/kg | $m^2 \cdot s^{-2}$ |
| dose equivalent[d] | sievert | Sv | J/kg | $m^2 \cdot s^{-2}$ |
| catalytic activity | katal | kat | | $s^{-1} \cdot mol$ |

[a] The radian and steradian may be used advantageously in expressions for derived units to distinguish between quantities of a different nature but of the same dimension;

[b] In practice, the symbols rad and sr are used where appropriate, but the derived unit "1" is generally omitted.

[c] In photometry, the unit name steradian and the unit symbol sr are usually retained in expressions for derived units.

[d] Other quantities expressed in sieverts are ambient dose equivalent, directional dose equivalent, personal dose equivalent, and organ dose equivalent.

**Table 4 Examples of SI derived units whose names and symbols include SI derived units with special names and symbols**

| Derived quantity | Name | Symbol |
|---|---|---|
| dynamic viscosity | pascal second | Pa · s |
| moment of force | newton meter | N · m |
| surface tension | newton per meter | N/m |
| angular velocity | radian per second | rad/s |
| angular acceleration | radian per second squared | $rad/s^2$ |
| heat flux density, irradiance | watt per square meter | $W/m^2$ |
| heat capacity, entropy | joule per kelvin | J/K |
| specific heat capacity, specific entropy | joule per kilogram kelvin | J/(kg · K) |
| specific energy | joule per kilogram | J/kg |
| thermal conductivity | watt per meter kelvin | W/(m · K) |
| energy density | joule per cubic meter | $J/m^3$ |
| electric field strength | volt per meter | V/m |
| electric charge density | coulomb per cubic meter | $C/m^3$ |
| electric flux density | coulomb per square meter | $C/m^2$ |
| permittivity | farad per meter | F/m |

continued

| Derived quantity | Name | Symbol |
|---|---|---|
| permeability | henry per meter | H/m |
| molar energy | joule per mole | J/mol |
| molar entropy, molar heat capacity | joule per mole kelvin | J/(mol · K) |
| exposure (x and $\gamma$ rays) | coulomb per kilogram | C/kg |
| absorbed dose rate | gray per second | Gy/s |
| radiant intensity | watt per steradian | W/sr |
| radiance | watt per square meter steradian | W/($m^2$ · sr) |
| catalytic (activity) concentration | katal per cubic meter | kat/$m^3$ |

**Table 5 SI prefixes**

| Factor | Name | Symbol | Factor | Name | Symbol |
|---|---|---|---|---|---|
| $10^{24}$ | yotta | Y | $10^{-1}$ | deci | d |
| $10^{21}$ | zetta | Z | $10^{-2}$ | centi | c |
| $10^{18}$ | exa | E | $10^{-3}$ | milli | m |
| $10^{15}$ | peta | P | $10^{-6}$ | micro | $\mu$ |
| $10^{12}$ | tera | T | $10^{-9}$ | nano | n |
| $10^{9}$ | giga | G | $10^{-12}$ | pico | p |
| $10^{6}$ | mega | M | $10^{-15}$ | femto | f |
| $10^{3}$ | kilo | k | $10^{-18}$ | atto | a |
| $10^{2}$ | hecto | h | $10^{-21}$ | zepto | z |
| $10^{1}$ | deka | da | $10^{-24}$ | yocto | y |

**Table 6 Units outside the SI that are accepted for use with the SI**

| Name | Symbol | Value in SI units |
|---|---|---|
| minute (time) | min | 1 min=60 s |
| hour | h | 1 h=60 min=3600 s |
| day | d | 1 d=24 h=86 400 s |
| degree (angle) | ° | $1°=(\pi/180)$ rad |
| minute (angle) | ′ | $1'=(1/60)°=(\pi/10\ 800)$ rad |
| second (angle) | ″ | $1''=(1/60)'=(\pi/648\ 000)$ rad |
| liter | L | $1\ \text{L}=1\ \text{dm}^3=10^{-3}\ \text{m}^3$ |
| metric ton[a] | t | $1\ \text{t}=10^3$ kg |
| neper | Np | 1 Np=1 |
| bel[b] | B | 1 B=(1/2) ln 10 Np[c] |
| electronvolt[d] | eV | $1\ \text{eV}=1.602\ 18\times10^{-19}$ J, approximately |
| unified atomic mass unit[e] | u | $1\ \text{u}=1.660\ 54\times10^{-27}$ kg, approximately |
| astronomical unit[f] | ua | $1\ \text{ua}=1.495\ 98\times10^{11}$ m, approximately |

[a] In many countries, this unit is called "tonne."

[b] The bel is most commonly used with the SI prefix deci: 1 dB=0.1 B.

[c] Although the neper is coherent with SI units and is accepted by the CIPM, it has not been adopted by the General Conference on Weights and Measures (CGPM, *Conférence Générale des Poids et Mesures*) and is thus not an SI unit.

[d] The electronvolt is the kinetic energy acquired by an electron passing through a potential difference of 1 V in vacuum. The value must be obtained by experiment, and is therefore not known exactly.

[e] The unified atomic mass unit is equal to 1/12 the mass of an unbound atom of the nuclide $^{12}C$, at rest and in its ground state. The value must be obtained by experiment, and is therefore not known exactly.

[f] The astronomical unit is a unit of length. Its value is such that, when used to describe the motion of bodies in the solar system, the heliocentric gravitation constant is $(0.017\ 202\ 098\ 95)^2\ \text{ua}^3\cdot\text{d}^{-2}$. The value must be obtained by experiment, and is therefore not known exactly.

**Table 7 Other units outside the SI that are currently accepted for use with the SI, subject to further review**

| Name | Symbol | Value in SI units |
|---|---|---|
| nautical mile | | 1 nautical mile = 1852 m |
| knot | | 1 nautical mile per hour = (1852/3600) m/s |
| are | a | 1 a = 1 $dam^2 = 10^2$ $m^2$ |
| hectare | ha | 1 ha = 1 $hm^2 = 10^4$ $m^2$ |
| bar | bar | 1 bar = 0.1 MPa = 100 kPa = 1000 hPa = $10^5$ Pa |
| ångström | Å | 1 Å = 0.1 nm = $10^{-10}$ m |
| barn | b | 1 b = 100 $fm^2 = 10^{-28}$ $m^2$ |
| curie | Ci | 1 Ci = $3.7 \times 10^{10}$ Bq |
| roentgen | R | 1 R = $2.58 \times 10^{-4}$ C/kg |
| rad | rad | 1 rad = 1 cGy = $10^{-2}$ Gy |
| rem | rem | 1 rem = 1 cSv = $10^{-2}$ Sv |

# References

[1] 邱春蓉，黄整. 大学物理实验双语教程. 成都：西南交通大学出版社，2010.

[2] Bernard C H, Epp C D. Laboratory Experiments in College Physics. 5th ed. New York: John Wiley & Sons, 1980.

[3] Wilson Jerry D. Hwenandez Cexilia A. Physics Labortory Experiments. 6th ed. Boston: Houghton Mifflin Company, 2003.

[4] 唐晋生，吴宗森，盛克敏. 大学物理实验(双语教学用书). 北京：国防工业出版社，2011.

[5] Halliday David, Resnick Robert, Walker Jearl. Fundamentals of Physics, 6th ed. New York: John Wiley & Sons, 2003.

[6] Wilson Jerry D. Physics Laboratory Experiments. 5th ed. Boston: Houghton Mifflin Company, 1998.

[7] Preston D W. Experiments in Physics. New York: John Wiley & Sons, 1985.